SpringerBriefs in Applied Sciences and Technology

Computational Intelligence

Series Editor

Janusz Kacprzyk

For further volumes:
http://www.springer.com/series/10618

Frederico A. E. Rocha · Ricardo M. F. Martins
Nuno C. C. Lourenço · Nuno C. G. Horta

Electronic Design Automation of Analog ICs Combining Gradient Models with Multi-Objective Evolutionary Algorithms

Frederico A. E. Rocha
Instituto de Telecomunicações
Instituto Superior Técnico
Lisbon
Portugal

Ricardo M. F. Martins
Instituto de Telecomunicações
Instituto Superior Técnico
Lisbon
Portugal

Nuno C. C. Lourenço
Instituto de Telecomunicações
Instituto Superior Técnico
Lisbon
Portugal

Nuno C. G. Horta
Instituto de Telecomunicações
Instituto Superior Técnico
Lisbon
Portugal

ISSN 2191-530X   ISSN 2191-5318   (electronic)
ISBN 978-3-319-02188-1   ISBN 978-3-319-02189-8   (eBook)
DOI 10.1007/978-3-319-02189-8
Springer Cham Heidelberg New York Dordrecht London

Library of Congress Control Number: 2013947787

Printed on acid-free paper

Springer is part of Springer Science+Business Media (www.springer.com)

To my parents and Susana

Frederico Rocha

To Nádia and Daniela

Ricardo Martins

To Alina

Nuno Lourenço

To Carla, João and Tiago

Nuno Horta

Preface

In the last years, the world has observed the increasing complexity of integrated circuits (ICs), strongly triggered by the proliferation of consumer electronic devices. The design of complex system on a chip (SoC) is widespread in multimedia and communication applications, where the analog and mixed-signal (AMS) blocks are integrated together with digital circuitry. However, the analog blocks development cycles are larger when compared to the digital counterpart. The two main reasons identified are the lack of effective computer-aided-design (CAD) tools for electronic design automation (EDA), and that analog circuits are being integrated using technologies optimized for digital circuits. Given the economic pressure for high-quality yet cheap electronics and challenging time-to-market constraints, there is an urgent need for CAD tools that increase the analog designers' productivity and improve the quality of resulting ICs.

The work presented in this book belongs to the scientific area of electronic design automation and addresses the circuit-level sizing and optimization of analog ICs. Particularly, an innovative approach to enhance a state-of-the-art layout-aware analog IC circuit-level optimizer, by embedding statistical knowledge from an automatically generated gradient model into the multi-objective multi-constraint optimization kernel based on a modified NSGA-II algorithm. The gradient model is automatically generated by, first, using a design of experiments (DOE) approach with two alternative sampling strategies, the full factorial design and the fractional factorial design, which define the samples that will be accurately evaluated using a circuit simulator (e.g., HSPICE®), second, extracting and ranking the contributions of each design variable to each performance measure or objective, and, finally, building the model based on series of gradient rules. The gradient model is then embedded into the modified NSGA-II optimization kernel, by acting on the mutation operator. The approach was validated with typical analog circuit structures for an industry standard 0.13 μm integration process, showing that, by enhancing the circuit sizing evolutionary kernel with the gradient model, the optimal solutions are achieved, considerably, faster and with identical or superior accuracy.

The book is organized into six chapters.

Chapter 1 gives a brief introduction to the area of analog IC design automation, with special emphasis to the design flow hierarchy and the circuit-level sizing and optimization.

Chapter 2 presents an extensive state-of-the-art review on analog integrated circuit (IC) design automation tools applied to the circuit-level synthesis problem. Particularly, several circuit-level sizing techniques are sketched and compared, and then, different model-based optimization approaches are outlined.

Chapter 3 illustrates the Gradient Model generation. The circuit is first sampled using either the full factorial or the fractional factorial Design of Experiments (DOE) techniques, and then the main effect is used to extract the gradient rules which compose the Gradient Model.

Chapter 4 describes how the Gradient Model is used to enhance the circuit-level optimization tool, GENOM-POF. GENOM-POF is part of the Analog Integrated circuit Design Automation environment (AIDA), developed in the Integrated Circuits Group at Instituto de Telecomunicações, Lisboa, Portugal. The integration of the gradient model includes both embedding the model in the optimization kernel, and add the model's setup options to AIDA's graphical user interface (GUI), which allows the visualization of the results and the configuration of the parameters, such as the objectives, constraints and input variables, ranges, etc.

Chapter 5 illustrates the application of the proposed methodology to practical examples. The framework of the proposed methodology for the automatic generation of analog ICs layout has been coded in JAVA and is running, for the presented examples, on an Intel® Core™ 2 Quad CPU 2.4 GHz with 6 GB of RAM.

Chapter 6 summarizes the provided book and supplies the respective conclusion and future work.

Frederico A. E. Rocha
Ricardo M. F. Martins
Nuno C. C. Lourenço
Nuno C. G. Horta

Contents

Abbreviations

AMS	Analog and Mixed-Signal
CAD	Computer Aided Design
CMOS	Complementary Metal-Oxide-Semiconductor
DOE	Design of Experiments
DSP	Digital Signal Processing
EDA	Electronic Design Automation
FFNN	Feed Forward Neural Networks
GA	Genetic Algorithm
GP	Geometrical Programming
GUI	Graphical User Interface
IC	Integrated Circuit
MARS	Multivariate Adaptive Regression Splines
NSGA	Nondominated Sorting Genetic Algorithm
PRSA	Parallel Re-combinative Simulated Annealing
POF	Pareto Optimal Front
PVT	Process Voltage Temperature
RF	Radio Frequency
SA	Simulated Annealing
SoC	System-on-a-Chip
SVM	Support Vector Machine
VLSI	Very Large-Scale Integration

Chapter 1
Introduction

Abstract This chapter presents a brief introduction to analog integrated circuits (ICs) design and to the area of analog IC design automation. First, the analog IC design problem is presented, that led to the research in this area, then, the traditional analog design flow is sketched and, finally, the features of the proposed methodology to enhance the circuit-sizing task are outlined.

Keywords Analog IC design · Circuit sizing · Gradient rules · Electronic design automation · Computer-aided-design

1.1 Analog IC Design

In the last decades, Very Large Scale Integration (VLSI) technologies have been widely improved, allowing the proliferation of consumer electronics and enabling the growth of integrated circuits (IC) market from $10 billion in 1980 to over $300 billion in 2013 [1]. IC designers are building systems that are increasingly more complex and integrated. The need of new functionalities, smaller devices, longer battery life, e.g., more power efficiency, less production and integration costs, and less design cost makes the design of electronic systems a truly challenging task, which must be completed within strict time-to-market constraints.

Although most of the functionalities in a modern electronic system are implemented using digital and digital signal processing (DSP) circuitry, analog and radio frequency (RF) circuitry, being essentially the link between digital circuitry and the continuous-valued external world, is integrated in the same chip. In such systems on a chip (SoC), the analog part occupies only about 10 % of the circuit area, however, the development time of analog blocks is considerably higher when compared to the development time of the digital part. The three main reasons identified for the larger development time of analog blocks are: the lack of effective Computer Aided Design (CAD) tools for Electronic Design Automation (EDA); analog circuits are being integrated using technologies optimized for

F. A. E. Rocha et al., *Electronic Design Automation of Analog ICs Combining Gradient Models with Multi-Objective Evolutionary Algorithms*, SpringerBriefs in Computational Intelligence, DOI: 10.1007/978-3-319-02189-8_1, © The Author(s) 2014

digital circuits; and, analog blocks are difficult to reuse because they are more sensitive to environmental and process variations than its digital counterpart [2].

In digital IC design, several EDA tools and design methodologies are available that help the designers keeping up with the new capabilities offered by the technology processes. By its part, electrical simulators are the only analog design automation tool really established, despite the algorithms and techniques introduced in the last 25 years [3]. Due to the lack of automation, designers keep exploring the solution space almost manually. This method causes long design cycles, and allied to the non-reusable nature of analog IC, makes analog IC design a cumbersome task.

Designers have been replacing functions of analog circuits for digital processing whenever possible; however, there are some typical blocks that are appointed as remaining forever analog, such as [4]:

- On the input side of a system, the signals of a sensor, microphone or antenna have to be detected or received, amplified and filtered, to enable digitalization with good signal-to-noise and distortion ratio. Typical applications of these circuits are in sensor interfaces, telecommunication receivers or sound recording;
- Mixed-signal circuits like sample-and-hold, analog-to-digital converters, phase-locked loops and frequency synthesizers. These blocks provide the interface between the input/output of a system and digital processing parts of a SoC;
- On the output side of a system, the signal from digital processing must be converted and strengthened to analog so that the signal achieves the output with low distortion;
- Voltage/current reference circuits and crystal oscillators offer stable and absolute references for the sample-and-hold, analog-to-digital converters, phase-locked loops and frequency synthesizers;

The developments on the IC industry enabled the design of extremely complex Analog and Mixed-Signal (AMS) systems, which are established in telecommunications, medical and multimedia applications. To increase the performance of the ICs, i.e., enhance the functionalities but with lower power consumption, there is an exponential increase in the number of devices contained in a IC, as described by Moore's law. This means that the designers deal with the IC projects containing billions of transistors, under extreme competitive market conditions.

Despite the developments in the recent years, analog design automation tools and methodologies are still far from achieving a mature state, as there is no automation tool really established to support the analog design flow. Today's analog design is supported by circuit simulators, layout editing environments and verification tools, however the design cycle for AMS ICs is still long and error-prone.

In order to understand the automation of analog IC design, the steps in the design flow must be clear. After this brief introduction to the analog IC design problem, the systematic approach to the analog design automation flow [4], which intends to ease design automation, is covered in the next section.

1.2 The Analog IC Design Automation Flow

A typical and well accepted design flow for AMS ICs is presented in Fig. 1.1. This design flow consists of a series of top-down topology selection and specification translation steps, repeated from system level to the device level, and bottom-up layout generation, extraction and verification steps. Adopting a hierarchical top-down design methodology is possible to perform system architectural exploration, obtaining a better overall system optimization at a higher abstraction level before starting more detailed implementations at the device level. Thus, problems are found early in the design flow and, as a result, design have a higher chance of first-time success, with fewer or no overall time consuming redesign iterations.

On the top-down path, the topology selection is the process where a set of blocks and the connections between them in defined in order to implement the input specifications of the current hierarchy level. In the specification translation task, the higher-level specifications are translated in the specifications for each of the blocks. Block specifications may be the definition of the Gain and bandwidth for an amplifier, or the sizes of the transistors, depending of the models used in that abstraction level. The sizing is then verified to ensure the fulfillment of the input specifications.

At this point, the bottom-up flow is executed. Layout generation consists of creating the geometrical layout of the block under design at the lowest level in the

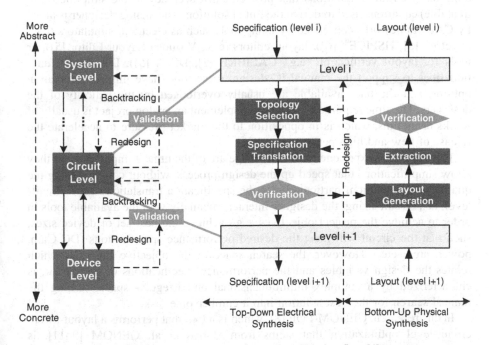

Fig. 1.1 System-level to device-level tasks of the analog IC design flow [4]

design hierarchy, or place and route the layouts of the sub-blocks at higher levels. Typically, the desired layout for a circuit is the one that minimizes the total area, while reducing the parasitic effects in the circuit performance. Then, the layout needs to be verified, which is done with design rule checkers and the layout-versus-schematic tools. Finally, the layout parasitics are extracted and simulated to verify its impact on the overall performance of the circuit.

The ascension to higher hierarchical levels is done when no potential problems are detected at the lowest levels and the layout meet the target requirements. When the topmost level verification is complete, the system is designed and ready for fabrication.

1.3 Research Contributions

This work addresses the problem of automatic specification translation at circuit level, also known as circuit sizing, where from the set of specifications, the designer finds out the sizes for the components, e.g., widths and lengths of the transistors, resistors, capacitors, etc. In the industry, this task is commonly done manually. The designers start by finding an approximate solution using simplified analytical expressions, and then, iteratively, adjust the solution until it meets all specifications, which sometimes can be very time consuming. The verification is done using circuit simulations that provide extra accuracy to the simplified (but treatable) equations used to derive the initial solution. The analog designer is aided by CAD frameworks comprised by many tools such as electrical simulators (e.g., Spectre® [5], HSPICE® [6]), layout editors (e.g., Virtuoso Layout Editor [5]), or tools for layout verification (e.g., CALIBRE [7], DIVA [5]). Despite its functionalities to support the manual IC design, these tools have limited automation options, and the ones available are usually overlooked by the majority of the designers. The time required to manually implement an analog project is usually of weeks or months, which is in opposition to the market pressure to accelerate the release of new and high performance ICs.

The designer's experience and knowledge are of the utmost importance, as they allow simplifications that speed up the design process, without compromising the quality of the solution, particularly, in the specification translation at the circuit-level, i.e., circuit sizing, the designer interacts manually with the available tools in order to achieve the project objectives, e.g., achieve the best set of device sizes, such that the circuit will meet the desired performance specifications (DC Gain, power, area, etc.). However, the search space of the objective function, which relates the design variables and the performance specifications of the circuit, is characterized by a complex multidimensional and irregular space, turning the manual search for the best solution into a cumbersome task.

In this research, GENOM-POF [8], which is a tool that performs a layout-aware circuit-level optimization that stems from Barros et al. GENOM [9–11], is enhanced by adding circuit specific knowledge that is automatically extracted using machine learning techniques. The circuit sizing is done using the Nondominated Sorting Genetic Algorithm (NSGA-II) [12] for multi-objective multi-constraint optimization, which addresses robust design requirements by considering Process Voltage Temperature (PVT) corner analysis, where Mentor Graphics® ELDO™ and Synopsys® HSPICE® circuit simulators are used for accurate evaluation of the circuit performance. This work aims to demonstrate the advantage of embedding simple statistical models, representing design knowledge, into the optimization kernel in order to improve the performance of the sizing optimization. The main objectives for this work are detailed below:

- Create a simple model that is capable of extracting a set of gradients rules, automated and autonomously, i.e., without any human knowledge. This set of gradients rules extracted should contain knowledge about any analog circuit in study;
- Create a model of rules and integrate it with the mutation operator of the (NSGA-II), in order to improve its efficiency during the optimization of the analog circuit. Compare the performance of reference NSGA-II with the modified NSGA-II with the model of gradients, created in the previous paragraph, and verify potential benefits of this modification;
- Evaluate and analyze the robustness of the models created previously, through its application in highly complex analog circuits;
- Improve the quality of the achieved sizing solutions.

The designer provides the chosen topology for the project, the variables for optimization and their ranges, the specifications to be met and the objective functions, e.g., minimize area/power, maximize DC Gain, etc., the tool instantiates the components to size, ensures that specifications are met and performs the search objectives space for the optimum solutions. The modified GENOM-POF, produced within this work, aims at helping the designer in his/her circuits sizing task, not only by generating solutions faster but also by achieving better Pareto optimal solutions.

1.4 Conclusions

The complexity of electronic systems imposes the use of CAD tools to support the design process. In digital IC design, several EDA tools and design methodologies are available that help the designers keeping up with the new capabilities offered by the technology, however the analog design automation tools strive to close the gap created due to the large investment made in the digital domain. This cause the manual exploration of the solution space, that in its turn creates expensively long designs that are difficult to reuse. In this context, the contributions of this research were presented, that aim to ease the efforts of analog designers to successfully complete this time-consuming task.

References

1. B. McClean, IC market to top \$300 billion for first time in 2013 (2011), [Online]. Available: http://www.icinsights.com
2. International Technology Roadmap for Semiconductors 2009 Edition (2009), [Online]. Available: http://public.itrs.net/
3. G.G.E. Gielen, CAD tools for embedded analogue circuits in mixed-signal integrated systems on chip. IEEE Proc. Comput. Digit. Tech, **152**(3), 317–332 (2005)
4. G.G.E. Gielen, R.A. Rutenbar, Computer-aided design of analog and mixed-signal integrated circuits. Proc. IEEE, **88**, 1825–1854 (2000)
5. Cadence Design Systems Inc, http://www.cadence.com
6. Synopsis, http://www.synopsys.com
7. Mentor Graphics, http://www.mentor.com
8. N. Lourenço, N. Horta, GENOM-POF: multi-objective evolutionary synthesis of analog ICs with corners validation, in *GECCO' 12: Proceedings of the fourteenth international conference on Genetic and evolutionary computation conference*, July 2012
9. M.F.M. Barros, J.M.C. Guilherme, N.C.G. Horta, *Analog circuits and systems optimization based on evolutionary computation techniques* (Springer, Berlin, 2010)
10. M. Barros, J. Guilherme, N. Horta, Analog circuits optimization based on evolutionary computation techniques, Integr. VLSI J, **43**(1), 136–155 (2010)
11. M. Barros, J. Guilherme, N. Horta, GA-SVM feasibility model and optimization kernel applied to analog IC design automation, in *Proceedings of ACM Great Lakes symposium on VLSI*, Stresa-Lago Maggiore, 2007
12. K. Deb, A. Pratap, S. Agarwal, T. Meyarivan, A fast and elitist multiobjective genetic algorithm: NSGA-II. Evol. Comput. IEEE Trans, **6**(2), 182–197 (2002)

Chapter 2
State-of-the-Art on Automatic Analog IC Sizing

Abstract In this chapter a state-of-the-art review on analog integrated circuit (IC) design automation tools applied to the specification translation problem is presented. Having the right topology for a given set of specifications is indispensable for a high performance design. An inadequate topology makes the design more difficult (or even impossible), and may require unnecessary resources, which is not acceptable in high performance designs. Once the topology is selected, the specifications for the overall block are translated to the specifications for the sub-blocks. The specifications are, in this way, passed through the hierarchy. At the lowest level, the translation reduces to circuit sizing, whereas at the higher levels it produce the sub-blocks performance parameters. In the last years, the scientific community proposed many techniques for the automation of the translation task; some apply only at circuit-level or only at system level, while others apply to both. In this study, several circuit-level sizing techniques are sketched and compared, and then, different model-based optimization approaches are outlined.

Keywords Analog IC design · Automatic specification translation · Knowledge-based sizing · Optimization-based sizing · Electronic design automation · Computer-aided-design

2.1 Automatic Circuit-Level Sizing

The techniques for the automation of circuit-level IC sizing are classified into two main groups [1], knowledge-based and optimization-based based on the techniques used to address the problem.

F. A. E. Rocha et al., *Electronic Design Automation of Analog ICs Combining Gradient Models with Multi-Objective Evolutionary Algorithms*, SpringerBriefs in Computational Intelligence, DOI: 10.1007/978-3-319-02189-8_2, © The Author(s) 2014

2.1.1 Knowledge-Based Sizing

Early strategies tried to systematize the design by using a design plan derived from expert knowledge. In these methods, a pre-designed plan is built with design equations and a design strategy that produce the component sizes that meet the performances requirements. Figure 2.1 shows the strategy flow of knowledge-based sizing methodologies.

In IDAC [2], the designer expertise is captured in a design plan where all design equations are explicitly solved during the execution of the plan. Once the topology is selected, the plan is executed for the given specifications to produce a first design. The tool also included local optimization around this first design. IDAC includes a vast library of plans, featuring voltage references, opAmps, comparators, oscillators, DACs and ADCs. OASYS [3] uses the same overall strategy, but defines the circuits hierarchically, with a design plan for each sub-block. It also adds backtracking with design-reuse methodologies to recover from failed designs. OASYS was extended to include data converters in addition to the original operational amplifiers. TAGUS [4–6] applies the design plan successfully at system-level for CMOS data converters. A slightly different approach is found in BLADES [7], CAMP [8] or ISAID [9, 10], these tools capture the designer's knowledge in expert systems using artificial intelligence techniques.

The knowledge-based approach was applied with moderate success. The main advantage of this approach is the short execution time. On the other hand, deriving the design plan is hard and time-consuming, the design plan requires constant maintenance in order to keep it up to date with technological evolution, and the results are not optimal, suitable only as a first-cut-design.

Fig. 2.1 Automatic circuit sizing: knowledge-based methodology

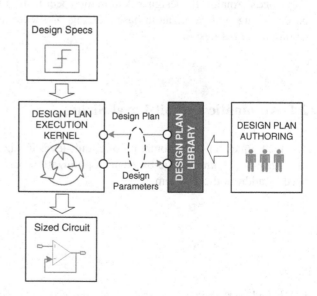

2.1.2 Optimization-Based Sizing

Aiming for optimality, the next generations of sizing tools apply optimization techniques to analog IC sizing. The optimization-based sizing can be classified into three major subclasses based on different techniques, namely, equation-based, simulation-based and model-based, which are addressed in the following subsections. A general flow of an optimization-based strategy can be found in Fig. 2.2.

2.1.2.1 Equation-Based

The equation-based methods use analytic design equations to evaluate the circuit performance. Different optimization techniques are used, the optimization in OPASYN [11] is done using steepest descent, whereas in STAIC [12] it is used a successive solution refinements technique. OPTIMAN [13] uses simulated annealing (SA) applied to analytical models created automatically by ISAAC [14]. DONALD [15] is an interactive design space exploration tool that assists the designer during circuit sizing by automatic analytical manipulations of the circuit equations. Maulik et al. [16] define the sizing problem as a constrained nonlinear optimization problem using spice models and DC operating point constraints, solving it using sequential quadratic programming. In ASTRX/OBLX [17] a simulated annealing optimization is performed using and cost function defined by equations for dc operation point, and small signal Asymptotic Waveform Evaluation based simulation. This evaluation technique is also used in DARWIN [18].

In GPCAD [19] a posynomial circuit model is optimized using Geometrical Programming (GP), the execution time is in the order of few seconds, but the

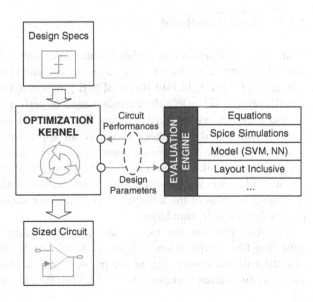

Fig. 2.2 Automatic circuit sizing: optimization-based methodology

general application of posynomial models is difficult and the time to derive the model for new circuits is still high. To reduce the long time spent in model development, automatic techniques were proposed (Gielen et al. in [20] provide a good overview on symbolic analysis applied to analog ICs). However, some design characteristics are still not easy to describe in analytical expressions with sufficient accuracy automatically. Kuo-Hsuan et al. [21] revisited the posynomial modeling recently, surpassing the accuracy issue by introducing an additional generation step, where local optimization using simulated annealing and a circuit simulator is performed. The same strategy is applied in FASY [22, 23] were analytical expressions are solved to generate an initial solution and a simulation-based optimization is performed to fine tune the solution.

The equation-based approaches are applied mostly at circuit-level, but some applications at system-level are also found. In SD-OPT [24] the optimal $\Delta\Sigma$ modulator sub-blocks' specifications are derived using symbolic equations solved using stochastic optimization. The sub-blocks itself are then generated using simulation based techniques. Doboli et al. [25] applies genetic programming techniques to simultaneously derive the sub-blocks specifications, sub-block topology selection and transistor sizing. Matsukawa et al. [26] design $\Delta\Sigma$ and pipeline analog to digital converters solving via convex optimization the equations that relate the performance of the converter to the size of the components.

The equation-based methods' strong point is the short evaluation time, making them, like the knowledge-based approaches, extremely suited to derive first-cut designs. The main drawback is that, despite the advances in symbolic analysis, not all design characteristics can be easily captured by analytic equations, in addition, the approximations introduced in the equations yield low accuracy designs especially for complex circuits.

2.1.2.2 Simulation-Based

With the availability of computing resources simulation based optimization gained ground. In simulation-based sizing a circuit simulator, like SPICE [27], is used to evaluate the circuit. In DELIGTH.SPICE [28] the optimization algorithm (phase I-II-III method of feasible directions) is used to perform local design optimization around a user provided starting point. Kuo-Hsuan et al. [21] and FASY [22, 23] use equation-based techniques to derive an approximate solution, and then use simulation within a simulated annealing optimization kernel to optimize the design. Cheng et al. [29] use the transistor bias conditions to constrain the problem and instead of solving the circuit by finding transistor sizes, the problem is solved by finding the bias of the transistors. The transistor sizes are derived from the bias point using electric simulation.

FRIDGE [30] on the other hand aims for global optimality by using an annealing-like optimization without any restriction to the starting point. However, to restrict the dimensionality of the problem the user still must provide the range for the optimization variables. In MAELSTROM [31] and ANACONDA [32] the evaluation time is reduced by a parallel mechanism that shares the evaluation load among multiple computers. Given the affinity evolutionary algorithms have with parallel implementations, it was the base technique chosen in MAELSTROM, however and because the success of simulated annealing is demonstrated in many implementations the authors option was to use parallel re-combinative simulated annealing (PRSA). In ANACONDA the approach is similar but instead of the PRSA it is applied a variation of pattern search algorithms, named by the authors as stochastic pattern search.

In order to account for layout induced effects and layout characteristics Castro-Lopez et al. [33] include the layout effects and parameters in the optimization. A template based layout generator is integrated in the optimization loop and the geometrical properties of the layout can be used as constraints or optimized. In addition layout parasitic are also extracted and used during the circuit's evaluation. They use simulated annealing followed by a deterministic method for fine-tuning to perform the optimization. The layout extraction is done using analytical equations and layout sampling or using 3-D geometric extraction models.

A different approach is taken in GENOM-POF [34], where a multi-objective strategy is applied through the use of evolutionary algorithms. The objectives and constraint functions are evaluated by HSPICE®. GENOM-POF outputs the Pareto optimal fronts (POF) with the tradeoff during the synthesis, so the designer has a wider range of solutions and choices to the problem of sizing.

Generality and easy-and-accurate model (the circuit netlist), are the strong points of simulation-based techniques. However, the execution time is large for complex circuits (~ 100 variables) and prohibitive at system level, and without the proper constraints the algorithm may not converge to a good result. Some heuristic schemes exist to automate the process of defining the constraints [35]. However, automatic constraint defining mechanisms are not integrated in sizing tools and their application is somewhat circuit class specific. Cheng et al. [29] uses manually derived DC point equations to limit the search space for the transistors dimensions.

Being the high execution time the weaker point of these methods, some techniques had been proposed to cope with it. Kuo-Hsuan et al. [21] used equation-based techniques to derive an approximate initial solution. Cheng et al. [29] instead of solving the circuit by finding transistor sizes, solved it by finding the bias of the transistors first, and then, the transistor sizes are derived from the bias point using electric simulation. In MAELSTROM [31] and ANACONDA [32] the evaluation time is reduced by a parallel mechanism that shares the evaluation load among multiple computers.

2.1.2.3 Model-Based

For some simulation-based approaches, macro models, like neural-networks or support vector machines (SVM), are also used to reduce the execution time caused by the use of circuit simulator in the loop. These models are automatically generated using an electric simulator to evaluate the performance of the training set.

Unlike the equations-based modeling the learning based modeling application to general circuits is easier; however, there is still the tradeoff between accuracy and model size and generation time.

Alpaydin et al. [36] use a neural-fuzzy model combined with an evolutionary optimization strategy where some of the AC performance metrics are computed using an equation-based approach. De Bernardinis et al. [37] use a learning tool based in SVMs to represent the performance space of analog circuits. The performance space is modeled using the knowledge acquired from a training set via circuit simulation.

Wolfe et al. [38] present a performance macro-model based in a neural network. This model once constructed, is to be used to replace the SPICE [27] simulation during the synthesis of analog circuits, increasing the efficiency of the performance parameter estimates' computation. The training and validation data sets are constructed with discrete points, sampled over the design space. The work explores several sampling methodologies to adaptively improve model quality and applies a sizing rules methodology in order to reduce the design space and ensure the correct operation of analog circuits.

Barros et al. [1, 39] present a cell-level synthesis and optimization approach based on SVMs and evolutionary strategies. The SVM is used to dynamically model performance space and identify the feasible design space regions while at the same time the evolutionary techniques are looking for the global optimum. The evaluation is still done with HSPICE® to ensure accuracy, but the number of evaluation is reduced by using the SVM to prune the candidate solutions.

A different approach is the use of POFs to explore circuit tradeoffs during synthesis [40], and instead of using a model for the circuits, the non-dominated solutions are generated (prior to the design task) and the suitable solution is selected from the already sized solutions. In [41], hierarchically POFs are used to perform system-level sizing. The POF-based-design execution time is large if the setup time (the generation of the POFs) is considered, however with the correct models, the POFs can be generated in a context free manner making then suitable for reuse.

In Tables 2.1, 2.2 and 2.3 the several tools for analog sizing automation are summarized and, in Table 2.4, the specification translation tools based on the techniques applied are compared.

2.2 Motivation for Model-Based Optimization

According to McConaghy and Gielen [42], there is a great improvement on the efficiency of an optimization cycle for analog IC sizing using electrical simulators, if models containing knowledge about the circuit are used. In [42] is presented a study to analyze the impact of different models in the optimization process, which were conducted for several different techniques: polynomials [43], posynomials [44], genetic programming [45], feedforward neural networks [46], boosted feedforward

Table 2.1 Overview of analog sizing tools, part I

Tool/author		Circuits	Design plan/optimization	Evaluation	Robust design	Topology gen.	Layout gen.	Time setup/ execution	Code
IDAC [2]	1987	Analog cells	Design plan plus SA post-optimization	Equations	⊘	✗	After sizing	Months/few sec	Pascal
DELIGTH.SPICE [28]	1988	Analog cells	Feasible directions optimization	SPICE-like	✓	✗	✗	Moderate/18 h	–
OASYS [3]	1989	OPAMP	Design plan (includes backtracking features)	Equations	✓	Before	✗	6 months/3 s	LISP
BLADES [7]	1989	OPAMP	Expert system for analog design	Equations	⊘	Before	✗	Long/20 min	LISP
OPASYN [11]	1990	OPAMP	Steepest descent	Equations	✓	Before	After sizing	2 weeks/5 min	C/LISP
CAMP [8]	1990	OPAMP	Expert system, flexible architecture	SPICE-like	⊘	During	After sizing	–/–	TURBO PROLOG
OPTIMAN [13]	1990	OPAMP	SA	Analytical models	✗	✗	✗	–/1 min	PASCAL
SEAS [52]	1991	OPAMP	SA	Equations	✗	During	✗	–/–	C
DONALD [15]	1991	OPAMP	Equation solver (Newton–Raphson variant)	Equations	⊘	✗	✗	–/–	LISP/ FORTAN
Chang [53]	1992	ADC	Top-down constraint driven	Behavior models	⊘	During	After sizing	–/4–89 h	C++
STAIC [12]	1992	OPAMP	2 step optimization	Equations	⊘	✗	After sizing	Long/2 min	C++
MINLP [54, 55]	1992	OPAMP	Branch & bound	Equations and BSIM models	⊘	During	✗	6 months/ 1 min	–

Table 2.2 Overview of analog sizing tools, part II

Tool/author		Circuits	Design plan/optimization	Evaluation	Robust design	Topology gen.	Latyout gen.	Time setup/ execution	Code
Maulik et al. [16]	1993	OPAMP	Sequential quadratic programming	Equations and BSIM models	⊘	✗	✗	6 months/ 1 min	C
FRIDGE [30]	1994	OPAMP	SA	SPICE-like	✗	✗	✗	1 h/45 min	–
DARWIN [18]	1995	OPAMP	Genetic Algorithm (GA)	Small signal, analytical expressions	✗	During	✗	–/–	–
ISAID [9, 10]	1995	OPAMP	Qualitative reasoning + post optimization	Equation and qualitative reasoning	⊘	✗	✗	–/–	C/PROLOG
SD-OPT [24]	1995	∑ Δ-modulator	SA	Equation adn behavioral simulation	✗	✗	✗	Long/ 1,5 week	–
FAST [22, 23]	1995	OPAMP	SA + Gradient	SPICE-like	✗	Before	✗	–/6 h	–
ASTRX/OBLX [17]	1996	Analog cells	SA	AWE equations	✗	✗	✗	few days/sec	C
Koza [56]	1997	Analog cells	GA	SPICE-like	✗	During	✗	–/–	C
GPCAD [57]	1998	OPAMP	Geometric programming	Posynomial models	✗	✗	✗	–/fast	MATLAB
Lohn [58]	1999	Filters	GA	SPICE-like	✗	During	✗	–/–	C
MAELSTROM [31]	1999	OPAM	GA + SA	SPICE-like	✗	✗	✗	–/3,6 h	C++
ANACONDA [32]	2000	OPAMP	Stochastic pattten search	SPICE-like	✗	✗	✗	–/10 h	C++
Sripramong [59]	2002	OPAMP	GA	SPICE-like	✗	During	✗	–/3 days	C

Table 2.3 Overview of analog sizing tools, part III

Tool/author		Circuits	Design plan/ optimization	Evaluation	Robust design	Topology gen.	Layout gen.	Time setup/ execution	Code
Alpaydin [36]	2003	OPAMP	Evolutionary strategies + SA	Fuzzy + NN trained with SPICE-Llike	✓	✗	✗	–/45 min	–
Shoou-Jin [60]	2006	Passive filters	GA	Equations	✗	During	✗	–/–	–
Barros [1, 39]	2006	Analog cells	GA	SPICE-like + feasibility SVM models	✓	✗	✗	–/20 min	C
Castro-Lopez [61]	2008	OPAMP	SA + Powels method	SPICE-like	⊘	✗	✓	–/25 min	–
MOJITO [62], [63]	2009	OPAMP	GP(NSGA-II)	SPICE-like	✓	During	✗	–/< 7 days	Python
Pradhan [64]	2009	OPAMP,filter	Multi-objective SA	Layout aware MNA models	✗	✗	✗	–/16 min	C ++
Matsukawa [26]	2009	ADC	Convex optimization	Convex functions	✓	After	✗	–/–	MATLAb
Cheng [29]	2009	OPAMP	SA	Equations	✗	✗	✓	–/< 1 h	C
Hongying [65]	2010	OPAMP	GA with VDE	SPICE-like	✗	During	✗	–/–	–
Kuo-Hsuan [21]	2011	RFDA	Covex optimization stochastic fine tuning	Posynomial SPICE-like	✗	✗	✗	–/1 h	MATLAB
GENOM-POF [34]	2012	OPAMP	Multi-objective GA	SPICE-like	✗	✗	✓	–/10 min	JAVA

Table 2.4 Classification of specification translation tools based on applied techniques and abstraction level

		Abstraction level	
		System-level	Cell-level
Knowledge-based		TAGUS [4–6] (+) Fast execution time (+) Use of expert knowledge (−) Expert knowledge is difficult to capture (−) not optimal	IDAC [2]; OASYS [3]; BLADES [7]; CAMP [8]; ISAID [9], [10] (+) Fast execution time (+) Use of expert knowledge (−) Expert knowledge is difficult to capture (−) not optimal
Optimization-based	Equation	SD-OPT [24]; Doboli [25]; Matsukawa [26] (+) Fast execution time (+) Use of expert knowledge (−) Difficult derivation of some equations (−) Simplifications lead to lack of accuracy	OPASYN [11]; STAIC [12]; Kuo-Hsuan [21]; OPTIMAN [13]; DONALD [15]; ASTRX/OBLX [17]; DARWIN [18]; GPCAD [57] (+) Fast execution time (+)* Use of expert knowledge (+)* Automatic symbolic analysis (−) Difficult derivation of some equations (−) Simplifications lead to lack of accuracy
	Simulation		Kuo-Hsuan [21]; FASY [22, 23]; ASTRX/OBLX [17]; DARWIN [18]; DELIGTH.SPICE [28]; Cheng [29]; FRIDGE [30]; MAELSTROM [31]; ANACONDA [32]; Castro-Lopez [61]; GENOM-POF [34] (+) Easy to develop models (++) Accurate and flexible (−) Still requires expert knowledge (−) Long execution time (−) Limited to cell-level
	Model		Alpaydin [36]; De Bernardinis [37]; Wolfe [38]; Barros [11], [39] (+) Accurate and flexible (−) Limited to cell-level

* not present in all approaches

neural networks [47], multivariate adaptive regression splines [48], support vector machines [49] and Kriging [50]. The choice of the models was based on their performance, and the following modeling methods were considered:

- As reference models were used: a constant (set as the mean of the data), a linear model and a 2nd-order polynomial;
- CAFFEINE [45] tool used a modified form of genetic programming (GP), which restricts GP to canonical function forms via a grammar;
- Feed forward neural networks (FFNNs) [46] which used the state-of-art training algorithm OLMAM;
- Boosting [47] creates a "stack" of models, each model is learned on a weighted version of the data. The overall output is the average of the outputs of the individual models;
- Multivariate Adaptive Regression Splines (MARS) [48] are piecewise polynomials. In the constructive steps, input variables are iteratively added on as "as-needed" basis for greedily chosen sub-regions of input space. MARS scales to a high number of input variables but is locally accurate;
- Support vector machines (SVMs) transform inputs into a space of much higher dimension and do linear regression in that space. A fast-learning variant LS-SVM [49] was used;
- Kriging [50] originated in geostatistics, but it has been shown to be useful in optimization. In this model prediction is the value of nearby samples "corrected" by a correlated error calculation.

Of the several existing ways to improve the optimization process efficiency, the study indicates that the construction of all models was based on the use of a Design of Experiments (DOE) technique [51].

Since electrical simulation is the bottleneck of the simulator-in-loop techniques, improving efficiency roughly translates to reducing the number of simulations. For a proper comparison between different models, a point that must be taken into account it is the setup time, i.e., the time necessary to create the model, which generally produces a tradeoff between model performance (accuracy and/or range of applicability) and model setup time.

Table 2.5 presents a summary of the study for the different models. From Table 2.5, CAFEINE is the approach with the better performance concerning the prediction error, while the Polynomial approach has the worst.. Based on this study, it is fair to forecast that with the type of approach made in CAFFEINE available, it could replace the simulator in the loop of an optimization process. However, the setup time of this model is huge when compared to the remaining; a model that has a setup time higher than the overall execution time is a huge contradiction.

Table 2.5 Comparison between several models for sizing automation of ICs

Model	Date	Heuristics	Circuits	Simulator	Time setup/ execution	Lang.	Error prediction (%)
Polynomial [43]	2005	Polynomial	High-speed CMOS	SPICE	1–4 min/ <10 min	Matlab	82,6
Posynomial [44]	2002	Posynomial	OTA, 13 inputs		1–4 min/ <10 min	Matlab	61,7
CAFFEINE [45]	2005	Posynomial	and 6 outputs		12 h/ <10 min	Matlab	22,7
FFNNs [46]	2002	Neural networks			3,7 min/ <10 min	Matlab	41,7
Boosted FFNN [47]	2002	Neural networks			7 min/ <10 min	Matlab	43,2
MARS [48]	1991	Polynomial			5 min/ <10 min	Matlab	29,4
LS-SVM [49]	2002	Support vector machine			5 min/ <10 min	Matlab	45,9
Kriging [50]	1998	Geostatistics			5 min/ <10 min	Matlab	34,6

2.3 Conclusions

Despite the evolution verified in the high and low abstraction levels, both architecture's selection, sizing and layout optimization remain the focus of research in analog EDA methodologies. The industrial commercial tools follow closely the main trends in academia and R&D workgroups, focusing in the lower level of abstraction levels dealing with device sizing and layout description levels.

Although much has been accomplished in automatic design of analog circuits, the fact is that custom generators usable in industrial design environment are not available. In this survey, some of the most significant analog design automation tools for circuit-sizing were presented and analyzed to provide a better understanding of its advantages and shortcomings. The tools are classified according to the techniques used and the applicability to cell and (or) system level.

Particularly, the results of Sect. 2.2 present a real motivation for a model-based optimization. The opportunity to create a new and innovative model, with a good performance both in terms of accuracy and setup time, arises. In this work, the idea of acquire knowledge of a circuit and embedding it into the evolutionary optimization kernel is explored. However, the model is used to guide the optimization kernel in a more efficient search of the solution space rather than replacing the usage of the circuit simulator to evaluate the performance of the circuit. The methodology adopted is to automatically generate a model that estimates how move to better solutions during the optimization. Chapter 3, describes the Gradient Model introduced in this work, and how it is automatically generated using DOE with two alternatives strategies, the Full Factorial Design and the Fractional Factorial Design. The model is then integrated into the synthesis tool AIDA, as will be presented in Chap. 4, and the obtained results are shown in Chap. 5.

References

1. M.F.M. Barros, J.M.C. Guilherme, N.C.G. Horta, *Analog circuits and systems optimization based on evolutionary computation techniques* (Springer, Berlin, 2010)
2. M.G.R. Degrauwe, O. Nys, E. Dijkstra et al., IDAC: an interactive design tool for analog CMOS circuits. IEEE J. Solid-State Circuits **22**(6), 1106–1116 (1987)
3. R. Harjani, R.A. Rutenbar, L.R. Carley, OASYS: a framework for analog circuit synthesis. IEEE Trans. Comput.Aided Des. Integr. Circuits Syst. **8**(12), 1247–1266 (1989)
4. N.C. Horta, J.E. Franca, High-level data conversion synthesis by symbolic methods, in *Proceedings of the IEEE International Symposium on Circuits and Systems*, vol. 4 (1996), pp. 802–805
5. N. Horta, Analogue and mixed-signal systems topologies exploration using symbolic methods. Analog Integr. Circ. Sig. Process **31**(2), 161–176 (2002)
6. N.C. Horta, J.E. Franca, Algorithm-driven synthesis of data conversion architectures. IEEE Trans. Comput. Aided Des. Integr. Circuits Syst. **16**(10), 1116–1135 (1997)
7. F. El-Turky, E.E. Perry, BLADES: an artificial intelligence approach to analog circuit design. IEEE Trans. Comput. Aided Des. Integr. Circuits Syst. **8**(6), 680–692 (1989)
8. B.J. Sheu, J.C. Lee, A.H. Fung, Flexible architecture approach to knowledge-based analogue IC design. IEEE Proc. G Circuits Devices Syst. **137**(4), 266–274 (1990)
9. C.A. Makris, C. Toumazou, Analog IC design automation. II. Automated circuit correction by qualitative reasoning. IEEE Trans. Comput. Aided Des. Integr. Circuits Syst. **14**(2), 239–254 (1995)
10. C. Toumazou, C.A. Makris, Analog IC design automation. I. Automated circuit generation: new concepts and methods. IEEE Trans. Comput.Aided Des. Integr.Circuits Syst. **14**(2), 218–238 (1995)
11. H.Y. Koh, C.H. Sequin, P.R. Gray, OPASYN: a compiler for CMOS operational amplifiers. IEEE Trans. Comput. Aided Des. Integr. Circuits Syst. **9**(2), 113–125 (1990)
12. J.P. Harvey, M.I. Elmasry, B. Leung, STAIC: an interactive framework for synthesizing CMOS and BiCMOS analog circuits. IEEE Trans. Comput. Aided Des. Integr. Circuits Syst. **11**(11), 1402–1417 (1992)
13. G.G.E. Gielen, H.C.C. Walscharts, W.M.C. Sansen, Analog circuit design optimization based on symbolic simulation and simulated annealing. IEEE J. Solid-State Circuits **25**(3), 707–713 (1990)
14. G.G.E. Gielen, H.C.C. Walscharts, W.M.C. Sansen, ISAAC: a symbolic simulator for analog integrated circuits. IEEE J. Solid-State Circuits **24**(6), 1587–1597 (1989)
15. K. Swings, W. Sansen, DONALD: a workbench for interactive design space exploration and sizing of analog circuits, in *Proceedings of the European Conference on Design Automation*, (1991), pp. 475–479
16. P.C. Maulik, L.R. Carley, D.J. Allstot, Sizing of cell-level analog circuits using constrained optimization techniques. IEEE J. Solid-State Circuits **28**(3), 233–241 (1993)
17. E.S. Ochotta, R.A. Rutenbar, L.R. Carley, Synthesis of high-performance analog circuits in ASTRX/OBLX. IEEE Trans. Comput. Aided Des. Integr. Circuits Syst. **15**(3), 273–294 (1996)
18. W. Kruiskamp, D. Leenaerts, DARWIN: CMOS opamp synthesis by means of a genetic algorithm, in *Proceedings of the Design Automation Conference*, (1995), pp. 433–438

19. M. del Mar Hershenson, S.P. Boyd, T.H. Lee, GPCAD: a tool for CMOS op-amp synthesis, in *InternationBaal Conference on Computer-Aided Design, Digest of Technical Papers of the IEEE/ACM*, (1998), pp. 296–303

20. G. Gielen, P. Wambacq, W.M. Sansen, Symbolic analysis methods and applications for analog circuits: a tutorial overview. Proc. IEEE **82**(2), 680–692 (1994)

21. M. Kuo-Hsuan, P. Po-Cheng, C. Hung-Ming, Integrated hierarchical synthesis of analog/RF circuits with accurate performance mapping, in *Symposium on Quality Electronic Design (ISQED)*, (2011), pp. 1–8

22. A. Torralba, J. Chavez, L.G. Franquelo, FASY: a fuzzy-logic based tool for analog synthesis. IEEE Trans. Comput. Aided Des. Integr. Circuits Syst. **15**(7), 705–715 (1996)

23. A.J. Torralba, J. Chavez, L.G. Franquelo, Fuzzy-logic-based analog design tools. IEEE Micro **16**(4), 60–68 (1996)

24. F. Medeiro, B. Perez-Verdu, A. Rodriguez-Vazquez et al., A vertically integrated tool for automated design of Sigma&Delta modulators. IEEE J. Solid-State Circuits **30**(7), 762–772 (1995)

25. A. Doboli, N. Dhanwada, A. Nunez-Aldana et al., A two-layer library-based approach to synthesis of analog systems from VHDL-AMS specifications. ACM Trans. Des. Autom. Electron. Syst. **9**(2), 238–271 (2004)

26. K. Matsukawa, T. Morie, Y. Tokunaga et al., Design methods for pipeline delta-sigma A-to-D converters with convex optimization, in *Design Automation Conference*, (2009), pp. 690–695

27. L.W. Nagel, *SPICE2: A Computer Program to Simulate Semiconductor Circuits* (EECS Department, University of California, Berkeley, 1975)

28. W. Nye, D.C. Riley, A. Sangiovanni-Vincentelli et al., DELIGHT.SPICE: an optimization-based system for the design of integrated circuits. IEEE Trans. Comput. Aided Des. Integr. Circuits Syst. **7**(4), 501–519 (1988)

29. L. Cheng-Wu, S. Pin-Dai, S. Ya-Ting et al., A bias-driven approach for automated design of operational amplifiers., in *International Symposium on VLSI Design, Automation and Test*, (2009), pp. 118–121

30. F. Medeiro, F.V. Fernandez, R. Dominguez-Castro et al., A Statistical Optimization-based Approach For Automated Sizing Of Analog Cells., in *Conference on Computer-Aided Design*, (1994), pp. 594–597

31. M. Krasnicki, R. Phelps, R.A. Rutenbar et al., MAELSTROM: efficient simulation-based synthesis for custom analog cells, in *Design Automation Conference*, (1999), pp. 945–950

32. R. Phelps, M. Krasnicki, R.A. Rutenbar et al., Anaconda: simulation-based synthesis of analog circuits via stochastic pattern search. IEEE Trans. Comput. Aided Des. Integr. Circuits Syst. **19**(6), 703–717 (2000)

33. R. Castro-Lopez, O. Guerra, E. Roca, F. Fernandez, An integrated layout-synthesis approach for analog ICs. IEEE Trans. Comput. Aided Des. Integr. Circuits Syst. **27**(7), 1179–1189 (2008)

34. N. Lourenço, N. Horta, GENOM-POF: Multi-Objective evolutionary synthesis of analog ICs with corners validation, in *GECCO' 12: Proceedings of the Fourteenth International Conference on Genetic and Evolutionary Computation Conference*, (2012), pp. 1119–1126

35. T. Massier, H. Graeb, U. Schlichtmann, The sizing rules method for CMOS and bipolar analog integrated circuit synthesis. IEEE Trans. Comput. Aided Des. Integr. Circuits Syst. **27**(12), 2209–2222 (2008)

36. G. Alpaydin, S. Balkir, G. Dundar, An evolutionary approach to automatic synthesis of high-performance analog integrated circuits. IEEE Trans. Evol. Comput. **7**(3), 240–252 (2003)

37. F. De Bernardinis, M.I. Jordan, A. SangiovanniVincentelli, Support vector machines for analog circuit performance representation, in *Design Automation Conference*, (2003), pp. 964–969

38. G.A. Wolfe, Performance Macro-Modeling Techiniques for Fast Analog Circuit Synthesis, University of Cincinnati, 2004

39. M. Barros, J. Guilherme, N. Horta, GA-SVM optimization kernel applied to analog IC design automation., in *IEEE Internation Conference on Electronics*, (2006), pp. 486–489
40. R. Castro-Lopez, E. Roca, F.V. Fernandez, Multimode pareto fronts for design of reconfigurable analogue circuits. Electron. Lett. **45**(2), 95–96 (2009)
41. E. Denize, G. Dundar, Hierarchical performance estimation of analog blocks using pareto fronts. Ph.D., Research in Microelectronics and Electronics, 2010
42. T. McConaghy, G. Gielen, Analysis of simulation-driven numerical performance modeling techniques for application to analog circuit optimization, in *IEEE International Symposium on Circuits and Systems (ISCAS)*, (2005), pp.1298–1301
43. T. McConaghy, G. Gielen, Analysis of simulation-driven numerical performance modeling techniques for application to analog circuit optimization, in *IEEE International Symposium on Circuits and Systems (ISCAS)*, (2005), pp. 1298–1301
44. W. Daems, G. Gielen, W. Sansen, Simulation-based generation of posynomial performance models for sizing of analog integrated circuits. IEEE Trans. CAD **22**(5), 517–534 (2003)
45. T. McConaghy, T. Eecklelaert, G. Gielen, CAFFEINE: Template-free symbolic model generation of analog circuits via canonical form functions and genetic programming, in *Design, Automation and Teste in Europe* **2**, (2005), pp. 1082–1087
46. N. Ampazis, S.J. Perantonis, OLMAN neural networks toolbox for Matlab (2002), http://iit.demokritos.gr/~abazis/toolbox/
47. R.E. Schapire, The boosting approach to machine learning: an overview, in *MSRI Workshop on Nonlin. Estimation and Classification*, (2002)
48. J.H. Friedman, Multivariate adaptive regression splines. Ann. Stat **19**, 1–141 (1991)
49. H. Drucker, C.J.C. Burges, L. Kaufman, A. Smola, V. Vapnik, in *Adv. in Neural Information Processing Systems 9*, ed. by M.C. Mozer, J.I. Jordan, T. Petscbe.Support vector regression machines, (MIT Press, Cambridge, 1997), pp. 155–161
50. D.R. Jones, M. Schonlau, W.J. Welch, Efficient global optimization of expensive black-box functions. J. Glob. Opt **13**(4), 455–492 (1998)
51. D.C. Montgomery, *Design and Analysis of Experiments*, 5th edn. (John Wiley and Sons, New York, 2001)
52. Z.Q. Ning, T. Mouthaan, H. Wallinga, SEAS: a simulated evolution approach for analog circuit synthesis, in *Proceedings of the IEEE Custom Integrated Circuits Conference*, (1991) pp. 5.2.1–5.2.4
53. H. Chang, A. Sangiovanlli-Vincentelli, F. Balarin et al., A top-down, constraint-driven design methodology for analog integrated circuits, in *Proceedings of the IEEE Custom Integrated Circuits*, vol. 3–6 (1992), pp. 8.4.1–8.4.6
54. P.C. Maulik, L.R. Carley, R.A. Rutenbar, A mixed-integer nonlinear programming approach to analog circuit synthesis, in *Proceedings of Design Automation Conference*, (1992), pp. 698–703
55. P.C. Maulik, L.R. Carley, R.A. Rutenbar, Integer programming based topology selection of cell-level analog circuits. IEEE Trans. Comput. Aided Des. Integr. Circuits Syst. **14**(4), 401–412 (1995)
56. J.R. Koza, F.H. Bennett III, D. Andre et al., Automated synthesis of analog electrical circuits by means of genetic programming. IEEE Trans. Evol. Comput. **1**(2), 109–128 (1997)
57. M. Del Mar Hershenson, S. P. Boyd, T. H. Lee, GPCAD: a tool for CMOS op-amp synthesis, in *International Conference on Computer-Aided Design, Digest of Technical Papers of the IEEE/ACM,*, vol. 8–12 (1998), pp. 296–303
58. J.D. Lohn, S.P. Colombano, A circuit representation technique for automated circuit design. IEEE Trans. Evol. Comput. **3**(3), 205–219 (1999)
59. T. Sripramong, C. Toumazou, The invention of CMOS amplifiers using genetic programming and current-flow analysis. IEEE Trans. Comput. Aided Des. Integr. Circuits Syst. **21**(11), 1237–1252 (2002)
60. C. Shoou-Jinn, H. Hao-Sheng, S. Yan-Kuin, Automated passive filter synthesis using a novel tree representation and genetic programming. IEEE Trans. Evol. Comput. **10**(1), 93–100 (2006)

61. R. Castro-Lopez, O. Guerra, E. Roca et al., An Integrated Layout-Synthesis Approach for Analog ICs. IEEE Trans. Comput. Aided Des. Integr. Circuits Syst. **27**(7), 1179–1189 (2008)
62. T. McConaghy, P. Palmers, M. Steyaert et al., Trustworthy genetic programming-based synthesis of analog circuit topologies using hierarchical domain-specific building blocks. IEEE Trans. Evol. Comput. **99**, 1–14 (2011)
63. P. Palmers, T. McConnaghy, M. Steyaert et al., *Massively Multi-Topology Sizing of Analog Integrated Circuits* (Design, Automation and Teste in Europe, 2009), pp. 706–711
64. A. Pradhan, R. Vemuri, Efficient synthesis of a uniformly spread layout aware pareto surface for analog circuits, in *22nd International Conference on VLSI Design*, (2009), pp. 131–136
65. Y. Hongying, H. Jingsong, Evolutionary design of operational amplifier using variable-length differential evolution algorithm, in *International Conference on Computer Application and System Modeling (ICCASM)*, (2010), pp. V4-610–V4-614
66. E. Roca, R. Castro-Lopez, F.V. Fernandez, Hierarchical synthesis based on pareto-optimal fronts, *in European Conference on Circuit Theory and Design*, (2009), pp. 755–758

Chapter 3
Gradient Model Generation

Abstract This chapter illustrates the Gradient Model generation. The circuit is first sampled using either the Full Factorial or the Fractional Factorial Design of Experiments (DOE) techniques, and then the main effect is used to extract the gradient rules which compose the Gradient Model.

Keywords Analog IC design · Design of experiments · Full factorial design · Fractional factorial design · Gradient model

3.1 Overview of Design of Experiments (DOE)

DOE is a highly used technique, as suggested in [1], to project and study the effects on the output (or response variables), by varying the input (or factors). Moreover, using this technique it is possible to make a statistical study of the output responses with a low cost, i.e., less computational time. According to [2], the steps for the development of the DOE are:

1. Characterization of the problem;
2. Selection of the response variables;
3. Choice of factor, levels, and ranges;
4. Choice of experimental design;
5. Conducting the experiment;
6. Statistical analysis of the data;
7. Conclusions and recommendations.

The purpose of using DOE is to extract the maximum amount of system information with the smallest number of runs. Here, with DOE, the influence of the inputs on outputs will be studied in order to enhance the process of automatically generate the sizing of a circuit, based on a NSGA-II kernel [3].

The first step towards the use of a sampling technique is to recognize and describe the problem to be tested and identify which are the objectives of the

F. A. E. Rocha et al., *Electronic Design Automation of Analog ICs Combining Gradient Models with Multi-Objective Evolutionary Algorithms*, SpringerBriefs in Computational Intelligence, DOI: 10.1007/978-3-319-02189-8_3, © The Author(s) 2014

experiment. In this case, the problem is to identify how the component sizes of an electrical circuit influence its performance measures by extracting gradient relations in order to, finally, compose a gradient model. In the next step of the DOE, it is necessary to select which output parameters are relevant (performance measures), due to a variation of the input. The output parameters and the ranges, as well as, the choice of factor and levels are provided by the user through a graphical user interface (GUI). Note that step 2 and 3 can be done simultaneously, or in the reverse order. The selection of factor and levels must take into account that the amount of simulations given by (3.1), where B is the base of the matrix, p it the number of non-elementary variables and n the total number of variables, is needed to construct the DOE matrix.

$$Number\ of\ Simulations = B^{(n-p)} \qquad\qquad (3.1)$$

The base of the matrix corresponds to the number of samples of input variables, the number of non-elementary variables corresponds to the variables which don't have all the possible combinations of sampling with the others variables, and the number of input variables corresponds to the variables defined by the designer as circuit variables. On the opposite of the non-elementary variables, the elementary variables correspond to the variables which have in the DOE matrix all the sampling combinations between the DOE matrix.

There is a trade-off between the base of the matrix and the number of elementary variables with the number of simulations. On one hand, the increase of the base matrix and the number of elementary variables produces a more robust experience; on the other hand, it increases the cost of computing the solution by increasing the number of simulations. The effect of the variation of the base and number of elementary variables in the DOE matrix will be studied in a later section.

In this model, the samples will only be obtained through the Factorial Design and Fractional Factorial Design. Other commonly used types of experiment design are:

- The Latin square design [2];
- The Greco-Latin square design [2].

In summary, and no matter the approach used, it is intended to create a DOE matrix for an evaluation of the output based on the selected input variables, in order to generate the gradient model. This process will be exemplified in the following sections for the Full Factorial and Fractional Factorial designs. For simplicity, only 2 levels (B = 2) will be experienced for the DOE matrix base and considering three inputs and two outputs, these values can be changed without loss of generality. For large DOE matrixes, the simulation is split into blocks of 1024 points due to limitations in the interface with HSPICE®, and also the heap of Java Virtual Machine is a limitation in terms of the maximum DOE matrix size.

3.2 Design of Experiments with Full Factorial Design

This section describes the design of experiments using Full Factorial design. The steps for the development of DOE, presented in the last section, are grouped in 2 sub-sections. The first, groups the problem definition steps (1 and 2) and the generation steps (3, 4 and 5), while the second sub-section groups the analysis steps (6 and 7).

3.2.1 Characterization and Construction of the DOE Matrix

The problem characterization is addressed by the presentation of an electrical schematic, where the input variables of the circuit and their respective ranges are used to set the choice of factor, levels and ranges. The selection of the response is defined by the outputs of this circuit. This step serves also to choose the type of experimental design used, the Full Factorial Design, and finally, the experiment is conducted.

In order to illustrate the description, the differential amplifier circuit shown in Fig. 3.1, with three input variables and two outputs, is used. The ranges for the input variables are provided by the user and shown in Table 3.1. The input variables (W1, 2 and W3, 4 represent, respectively, the Ws (widths) of the transistors pairs (M1, M2) and (M3, M4)). The output variables, also provided by the user, are shown in Table 3.2, whose values are obtained from circuit simulation using HSPICE®. Notice that both the inputs and outputs are intentionally a subset of the overall design parameters, e.g., L1, 2 and L3, 4 are fixed values, the main idea is to prove that even with an extremely simple gradient model it is possible to improve the optimization kernel by embedding design knowledge into the automation loop.

Sampling these ranges with the DOE technique implies an association between the values of the input variables within the range and the DOE levels to construct the DOE matrix. If the ranges are changed, the values associated with the level change, forcing a resampling. For this example two points in the range, i.e., the DOE matrix base has a value of two, are considered. The two levels are defined as

Fig. 3.1 Differential amplifier schematics

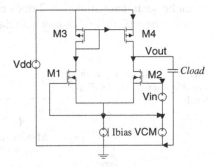

Table 3.1 Range of input variables

Inputs	Minimum range	Maximum range
W1, 2 (μm)	1	400
W3, 4 (μm)	1	400
IBias (μA)	100	500

Table 3.2 Objectives and design constraints

Outputs	Objective
DC gain (dB)	≥ 45
GBW (MHz)	≥ 35

low and high, described by 0 and 1 respectively (these are not logic levels). The range is divided into two equal parts, where the level 0 is associated to the middle of the lower half and level 1 to the middle of the upper half. For better understanding, the process is illustrated in Fig. 3.2. This kind of design is called the 2^k factorial design. In accordance with Montgomery [2], it is highly used in factor screening experiments, especially in systems where the response is approximately linear with the range of the factors. It is also a more simplified and fast design for a brief study of a system.

In the Full Fractional DOE the circuit is sampled in all the combinations of variables values. For each variable (x_i), B levels are defined and to each level a $v_{i,b}$ value, derived from the variable's range according to (3.2), is assigned.

$$v_{i,b} = X_i^{Min} + \frac{\left(X_i^{Max} - X_i^{Min}\right)}{2B} \times (1 + 2b), \qquad b = 0, \dots, B - 1 \qquad (3.2)$$

In Table 3.3 the mapping between level values and variable values is illustrated for the differential amplifier introduced previously, when B is set to 2.

Once the variables mapping is complete, the next step is to construct the DOE matrix. The matrix has one line per each possible combination of values, being the total number of lines given by (3.1), where p is 0. The columns are the inputs (x), identified with the levels, and the outputs (y) described by the measured values. The concatenation of the input's levels values is also referred as the code of the sample, as it acts as unique identifier. Table 3.4 shows the $8(2^3)$ sample matrix, obtained for the differential amplifier example. From the observation of the matrix, it can be seen that simulation 2 does not have values in the outputs. This situation occurs when HSPICE® cannot simulate the circuit, e.g. the simulation does not

Fig. 3.2 Relation between variable's range and DOE's levels

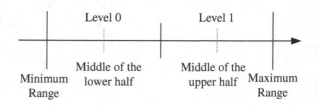

Table 3.3   Variables values for each level of the DOE

Variables	Level 0	Level 1
$x1 - W_{12}$ (µm)	100.75	300.25
$x2 - W_{34}$ (µm)	100.75	300.25
$x3 - IBias$ (µA)	200	400

Table 3.4   DOE matrix for Full Factorial design

	x_1 W_{12}	x_2 W_{34}	x_3 IBias	y_1 DC gain (dB)	y_2 GBW (MHz)
1	0	0	0	1.70	0.34
2	1	0	0	–	–
3	0	1	0	32.62	2.25
4	1	1	0	56.46	20.34
5	0	0	1	44.44	0.30
6	1	0	1	30.56	10.24
7	0	1	1	31.13	12.18
8	1	1	1	46.00	25.40

converge for that set of input parameters, or the circuit behavior renders the measurements of the outputs inoperative. All vectors which produce an output that is not measurable are not taken into consideration during the model generation.

For a better observation of space exploration performed through the DOE matrix, a hypercube is represented in Fig. 3.3 for the output DC Gain, however, this may be extrapolated to any other output. As can be seen through the hypercube, the greater the number of levels used in the DOE, the finer is the sampling of the search space, however, the tradeoff between search space and the execution time must be taken into account.

3.2.2 Analysis of the DOE Matrix

This phase presents the statistical analysis of the experiment conducted in the previous phase and the conclusions obtained from it. Having constructed the DOE matrix with the respective output values obtained from the circuit simulation, it is necessary to evaluate the effects of input variables on the outputs. This process is called the main effect of the input in the output.

The analysis of the data obtained in the DOE matrix, by calculating the main effect, is intended to identify which variable affects most each of the outputs. This conclusion is reached through the highest magnitudes of the main effect. The main effect is the effect of one independent (input) variable on the dependent (output) variable, ignoring the effects of all other independent variables.

Fig. 3.3 Full factorial (2^3)
hypercube for DC gain

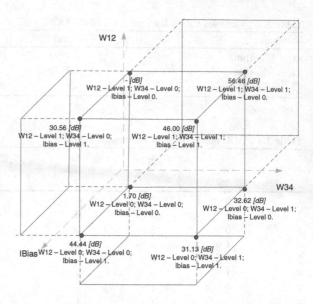

The Main Effect value is determined according to (3.3), where $m_{i,j}$, is the main effect of the input variable i in the output variable j, and k identifies the sample.

$$m_{i,j} = \sum_{k=1}^{B^{n-p}} w_{i,k} \times y_k, \qquad w_{i,k} = \begin{cases} +1 \ when \ x_{i,k} \geq \frac{B}{2} \\ -1 \ when \ x_{i,k} < \frac{B}{2} \end{cases} \qquad (3.3)$$

When the total Main Effect of an input variable is positive/negative, this is an indication that if the value of that input variable is increased, the value of the output will tend to increase/decrease. For the differential amplifier example, Table 3.5 shows the main effects of the input variables to the outputs (DC Gain and GBW) for the fully factorial DOE.

Table 3.5 Main effect obtained from the full factorial DOE matrix

Input	Effect on output y_1 − DC GAIN (dB)	$m_{i, 1}$
$x_1 - W_{12}$	(56.46 + 30.56 + 46.00) − (1.70 + 32.62 + 44.44 + 31.13)	23.13
$x_2 - W_{34}$	(32.62 + 56.46 + 31.13 + 46.00) − (1.70 + 44,44 + 30.56)	89.51
x_3 − Ibias	(44.44 + 30.56 + 31.13 + 46.00) − (1.70 + 32.62 + 56.46)	61.35

Input	Effect on output y_2 − DC GBW (MHz)	$m_{i, 2}$
$x_1 - W_{12}$	(20.34 + 10.24 + 25.40) − (0.34 + 2.25 + 0.30 + 12.18)	40.91
$x_2 - W_{34}$	(2.25 + 20.34 + 12.18 + 25.40) − (0.34 + 0.30 + 10.24)	49.29
x_3 − IBias	(0.30 + 10.24 + 12.18 + 25.4) − (0.34 + 2.25 + 20.34)	25.19

3.3 Design of Experiments with Fractional Factorial Design

The increase in the number of input variables leads, for the Full Factorial Design of experiments, to an exponential increase in the number of simulations, seen in (3.1), which, as mentioned previously, increases exponentially the time required to complete the whole DOE process. In order to attenuate this effect, the Fractional Factorial DOE introduces the notion of non-elementary variable, as a variable that is not used to generate the code of the sample, therefore reducing the size of the matrix, and the level of the non-elementary variables is determined from the code, i.e. from the levels of the elementary ones.

3.3.1 Characterization and Construction of the DOE Matrix

Fractional Factorial Design corresponds to increase the non-elementary variables in the construction of the matrix DOE, i.e., $p > 0$. Using $p = 1$, with $B = 2$ the total number of simulations obtained from (3.1) is $2^{3-1} = 4$. The number of simulations decreases by half in comparison with Full Factorial Design studied above. For the circuit in study, the reduction in the number of simulations is irrelevant. However, it is used as a demonstration for future use in more complex circuits. To illustrate the method, the variable IBias will be used as a non-elementary variable.

The level values of the two elementary variables are generated in the same way as in the Full Factorial Design. To compute the levels of the non-elementary variables, several methods are available in the literature [2], in this work the level, Ln_i, of the non-elementary variable i is given by (3.4), where mod is the modulo operator, and L_1 and L_2 are the levels of the first and second elementary variables, which ensures an even distribution in the levels.

$$Ln_i = (L_1 + L_2) \bmod B \qquad (3.4)$$

Table 3.6 shows the $4(2^{(3-1)})$ samples which compose the DOE matrix, obtained for the differential amplifier example by considering the variable IBias as a non-elementary variable.

3.3.2 Analysis of the DOE Matrix

After the construction of DOE matrix, the next step is to perform the statistical analysis of the data. This statistical study is made through the calculation of the main effect as performed in Sect. 3.2.2. Table 3.7 shows the main effects obtained

Table 3.6 DOE matrix for Fractional Factorial design

	x_1 W_{12}	x_2 W_{34}	x_3 IBias	y_1 DC gain (dB)	y_2 GBW (MHz)
1	0	0	0	1.70	0.34
2	1	0	1	30.56	10.24
3	0	1	1	31.13	12.18
4	1	1	0	56.46	20.34

Table 3.7 Main effect obtained from the fractional factorial DOE matrix

Input	Effect on output y_1 − DC gain (dB)	$m_{i,\,1}$
$x_1 - W_{12}$	$(30.56 + 56.46) - (1.70 + 31.13)$	54.19
$x_2 - W_{34}$	$(31.13 + 56.46) - (1.70 + 30.56)$	55.13
$x_3 - $ Ibias	$(30.56 + 31.13) - (1.70 + 56.46)$	3.53

Input	Effect on output y_2 − DC GBW (MHz)	$m_{i,\,2}$
$x_1 - W_{12}$	$(10.24 + 20.34) - (0.34 + 12.18)$	18.06
$x_2 - W_{34}$	$(12.18 + 20.34) - (0.34 + 10.24)$	21.94
$x_3 - $ IBias	$(10.24 + 12.18) - (0.34 + 20.34)$	1.74

from the fractional factorial DOE matrix. Both Tables 3.5 and 3.7 show all variables having a positive effect in the output, due to its highest absolute value in both cases; X2 is the input variable that gives more certainty in its effect towards both outputs. Observing the tables of Main Effects for both DOE strategies it is possible to conclude that both strategies have concordant directions.

3.4 Extraction of the Gradient Model from DOE

The generation of the model aims to assign a direction for both inputs and outputs, i.e., given a target variation of one of the outputs, it obtain the most likely variation of the inputs that lead to that outcome. This knowledge about the gradient of input variables and their respective response effect on the outputs is the final product that will be introduced into the optimizer.

Once the main effects are computed, the N input variables, where N is a parameter provided by the user, that have larger contributions to each output are identified as the ones with the large absolute Main Effect, and then, a refinement procedure is executed. For each output variable y_j, a new DOE matrix is constructed using the fractional factorial sampling, with the N input variables that have the larger contributions as the only elementary variables.

The refined DOE matrix is then converted to the set of gradient rules for that output variable. This is done by discarding the columns referring to non-elementary variables and transforming the levels of the elementary variables x_i

Table 3.8 Extraction of gradient rules for GBW

$y_{j,k}$	$Y_j^{Max} - \Delta_j$	$Y_j^{Min} + \Delta_j$	$So_{j,k}$
$y_{2,\,1} = 0.34$			$So_{2,1}$: $(-)$
$y_{2,\,2} = 10.24$	17.03	8.67	$So_{2,2}$: (u)
$y_{2,\,3} = 12.18$			$So_{2,3}$: (u)
$y_{2,\,4} = 20.34$			$So_{2,4}$: $(+)$

into input gradient symbols $si_{i,j,k}$ according to (3.5), where k identifies the line of the matrix.

$$si_{i,j,k} = \begin{cases} (+)\ when\ x_{i,k} \geq \frac{B}{2} \\ (-)\ when\ x_{i,k} < \frac{B}{2} \end{cases} \qquad (3.5)$$

The output gradient symbols $So_{j,k}$ are converted from the output values using (3.6), where Y_j^{Max} and Y_j^{Min} are respectively the maximum and minimum values of the output y_j obtained in the DOE matrix (not the refined matrix), and Δ_j is $\left| Y_j^{Max} - Y_j^{Min} \right| / 3$. The meanings of the symbols are: $(-)$ a decrease; $(+)$ increase and (u) undefined. Tables 3.8 and 3.9 illustrate the process of extracting Gradient Rules.

$$So_{j,k} = \begin{cases} (+)\ when\ y_{j,k} \geq Y_j^{Max} - \Delta_j \\ (u)\ when\ \left(Y_j^{Min} + \Delta_j\right) < y_{j,k} < \left(Y_j^{Max} - \Delta_j\right) \\ (-)\ when\ y_{j,k} \geq Y_j^{Max} - \Delta_j \end{cases} \qquad (3.6)$$

From the example of Table 3.9 some conclusions can be drawn. Both rules 2 and 3 do not give information about a gradient for the output, thus these rules cannot be applied. On the other hand, rules 1 and 4 show a gradient for the output, rule 1 should be invoked when one intends to decrease/minimize the value of this output, as set by the output gradient symbol $(-)$. This rule dictates that to achieve this variation in the output, both the values of W_{12} and W_{34} should be decreased. Similarly, rule 3 indicates that by increasing the values of W_{12} and W_{34}, the output value should increase/maximize, since it has the gradient symbol $(+)$.

Figure 3.4 summarizes the correspondence between the gradient symbols and the directions of the output variables.

Table 3.9 Set of gradient rules for GBW

	$si_{1,2}$ $W_{1\,2}$	$si_{2,2}$ $W_{3\,4}$	So_2 GBW
1	$(-)$	$(-)$	$(-)$
2	$(+)$	$(-)$	(U)
3	$(-)$	$(+)$	(U)
4	$(+)$	$(+)$	$(+)$

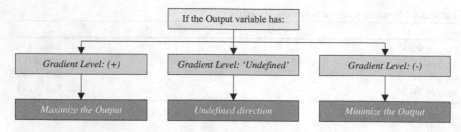

Fig. 3.4 Summary of the directions of the output variables

For better understanding of the Gradient Model generation the pseudo-code is presented in algorithm 3.1.

Algorithm 3.1 Gradient model generation procedure for n input variables, p non-elementary variables, and M outputs

1.	Set the parameters B, N
2.	Set P = the p non-elementary variables, E = the n elementary variables
3.	Set Rules = {}
4.	Set Matrix = fractionalFactorialSampling(B, E, P)
5.	for each output y_j : $j=1,…,M$ do
5.1	Y_j^{Max} = maxValueOfY(j, $Matrix$)
5.2	Y_j^{Min} = minValueOfY(j, $Matrix$)
5.3	Δ_j = (Y_j^{Max} - Y_j^{Min})/3
5.4	C_j = findInputsWithLargerMainEffect($Matrix$, j, N)
5.5	rMatrix$_j$ = fractionalSampling(B, C_j, $otherInputs$)
5.6	for each row k in rMatrix$_j$ do
5.6.1	Set rule$_{j,k}$ = new $emptyRule$
5.6.2	rule.so$_j$ = computeSo(j, k, Y_j^{Min}, Y_j^{Max}, Δ_j) ;$equation$ (3.6)
5.6.3	for each input x_i in C_j : $i=1,…,N$ do
	rule$_{j,k}$.si$_i$ = computeSi(i, k, B) ;$equation$ (3.5)
5.6.4	Set Rules = {rule$_{j,k}$} + $Rules$
6.	return Rules

Table 3.10 Overview of designs: Full Factorial and Fractional Factorial

Design	Full factorial	Fractional factorial
Advantages	Robust study of circuits	Lower runtime
	Considerer all selected variables of the circuit as elementary	Convenient for an early draft of the circuit
Disadvantages	High computational cost (for complex circuits)	Lower accuracy in the study of circuits
	High runtime (for complex circuits)	Difficulty in determining, which variables are non-elementary

3.5 Conclusions

In this chapter, two different techniques were demonstrated for the extraction of knowledge for analog ICs through the use of sampling, the Full Factorial Design and Fractional Factorial Design. Table 3.10 presents an overview of both techniques. Additionally, the calculation of the effect of the input variables on the output was also presented, this evaluation is extremely important to the extraction of the gradient rules, since it is the main indicator of the variables that most contribute to change the outputs as well as the direction of that change. Then, after the sampling process and its statistical study, the extraction of the Gradient Model step-by-step was presented. A refinement to the variables with larger main effect was carried out in order to obtain a more accurate model for these variables. Finally, it was explained how to generate the gradient rules for the input and output variables based on the refined matrix. The model is intended to be included in the optimization kernel to increase the performance, the process of integration is described in the next chapter.

References

1. J. Antony, N. Capon, Teaching experimental design techniques to industrial engineers. Int. J. Eng. **5**, 335–343 (1998)
2. D.C. Montgomery, *Design and Analysis of Experiments*, 5th edn. (Wiley, New York, 2001)
3. K. Deb, A. Pratap, S. Agarwal, T. Meyarivan, A fast and elitist multiobjective genetic algorithm: NSGA-II. IEEE Trans. Evol. Comput. **6**(2), 182–197 (2002)

Chapter 4
Enhanced AIDA's Circuit-Level Optimization Kernel

Abstract This chapter describes how the Gradient Model described in the previous chapter is used to enhance the circuit-level optimization tool, GENOM-POF [1]. GENOM-POF is part of the Analog Integrated circuit Design Automation environment (AIDA) [2], developed in the Integrated Circuits Group at Instituto de Telecomunicações, Lisboa, Portugal. The integration of the gradient model includes both embedding the model in the optimization kernel, and add the model's setup options to AIDA's graphical user interface (GUI), which allows the visualization of the results and the configuration of the parameters, such as the objectives, constraints and input variables, ranges, etc.

Keywords Analog IC design · Circuit-level sizing · Optimization-based sizing · Genetic algorithm · Genetic operators · Gradient model

4.1 Architecture

The AIDA platform [2], whose general architecture is shown in Fig. 4.1, implements a fully automatic approach from a circuit level specification to physical layout description. AIDA monitors the implemented design flow allowing the designer to intervene, e.g., by stopping the synthesis process whenever an acceptable solution is already achieved or by selecting the solution to be integrated from a Pareto set of optimally sized circuits.

The automatic analog IC design flow supported by AIDA consists of two phases. The first phase is the specification translation, or circuit sizing at circuit-level, where the sizes of the devices are determined in such way that the circuit fulfills the specifications. In AIDA, this task is performed by the circuit-level sizing tool, GENOM-POF [1]. The second phase is the physical implementation of the devices constrained to the technology design rules, which is generated by LAYGEN II [3–5].

F. A. E. Rocha et al., *Electronic Design Automation of Analog ICs Combining Gradient Models with Multi-Objective Evolutionary Algorithms*, SpringerBriefs in Computational Intelligence, DOI: 10.1007/978-3-319-02189-8_4, © The Author(s) 2014

Fig. 4.1 AIDA architecture

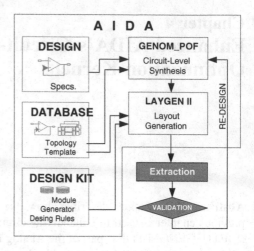

The contributions described in this book relate to GENOM-POF, it uses an optimization kernel based on adapted implementation of the multi-objective evolutionary algorithm NSGA-II [6] and the circuit simulator HSPICE®, to evaluate the design performance. Although GENOM-POF allows the inclusion of corner cases during optimization, this does not fall within the scope of this book, and is not addressed. The architecture of the GENOM-POF integrated with the developed Gradient Model is shown in Fig. 4.2.

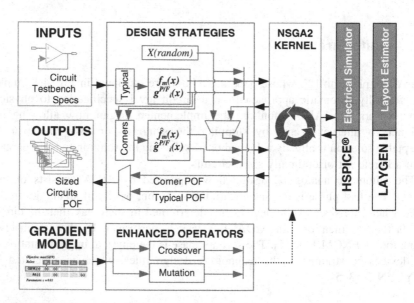

Fig. 4.2 GENOM-POF architecture with the gradient model integration

Fig. 4.3 Schematic of the single-ended folded cascode differential amplifier

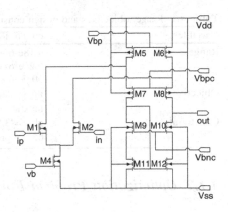

As inputs, GENOM-POF receives the circuit net list and test bench. These two files, provided by the designer, describe the circuit and specify both the optimization variables and the design objectives and constraints. The circuit is then modeled as an optimization problem suitable to be optimized by the NSGA-II kernel. As show in Fig. 4.2, the gradient model is used to modify the optimization kernel operators, using the direction defined in the model to affect the new solutions generated by the operators. The main idea is to accelerate the convergence process by embedding some knowledge of the circuit in the optimization kernel.

The output is a set of Pareto Optimal Fronts (POFs) with different sizing solutions, presenting the tradeoff between the objectives being optimized. From these outputs, the designer selects the ones to be used in the automatic layout generator tool, LAYGEN II, where the complete physical design is executed.

In the following sections, the single-ended folded cascode differential amplifier, illustrated in Fig. 4.3, will be used to exemplify the proposed automatic design flow.

4.1.1 Inputs

The inputs are provided by the designer, and consist in the netlist of the circuit and test benches in the HSPICE® format. This netlist should contain both the parameterization of the optimization variables and the measurement statements for the performance outputs.

The designer also has to define: the ranges of the optimization variables; the design constraints; and the optimization objectives. Table 4.1 presents a possible configuration that can be introduced as input for the single-ended folded cascade example.

Table 4.1 Range, objectives and design constraints example

Variables	cn, cp, ib, l1, l4, l5, l7, l9, l11, w1, w4, w5, w7, w9, w11
Ranges	0.18e-6 <= l* <= 5.0e-6
	0.24e-6 <= w* <= 200.0e-6
	30.0e-6 <= ib <= 400.0e-6
Objectives	min(area)
	max(a0)
Constraints	gb >= 3.5e7
	65 <= pm <= 90

4.1.2 Optimization Problem Formulation

GENOM-POF optimization engine is based on the NSGA-II algorithm, modified to interface with the circuit simulator, which performs the evaluation of each potential solution, i.e., population element. The reason why the NSGA-II was chosen over other multi-objective evolutionary algorithms was due to its excellent characteristics to produce Pareto optimal fronts, as mentioned in [4]. Regarding the choice of the circuit simulator to evaluate the performance of the circuits, HSCPICE®, was chosen because of its high accuracy. The modified GENOM-POF uses the same structure and the model is integrated into the evolutionary operators: crossover and mutation. The integration of the model into the tool GENOM-POF and its interface with the designer are explained in detail throughout the rest of this chapter.

The NSGA II multi-objective optimization kernel was designed to solve the problem defined in (4.1) where, x is a vector of N optimization variables, $f_m(x)$ is a set of M objective functions to minimize, $g_j(x)$ is the set of J constraints to be met and, finally, $x_i^L \leq x_i \leq x_i^U$ is the range of the variable x_i to be optimized.

$$\begin{aligned} find\ x\ that\ minimize\ f_m(x) \quad & m = 1, 2, \ldots M \\ subject\ to\ g_j(x) \geq 0 \quad & j = 1, 2, \ldots J \\ x_i^L \leq x_i \leq x_i^U \quad & i = 1, 2, \ldots N \end{aligned} \quad (4.1)$$

Thus, the first step to apply it to the circuit synthesis problem involves transforming the design problem in an optimization problem that may be executed by the NSGA-II kernel. The design objectives being minimized are used directly as one $f_m(x)$, and the ones being maximized are multiplied by -1. The design constraints are normalized and multiplied by -1, if necessary, according to (4.2), where, p_j is the measured circuit characteristic, and P_j is the corresponding acceptable limit. Table 4.2 illustrates the objective and constraint functions for the

Table 4.2 $f_m(x)$ and $g_j(x)$ for the example from Fig. 4.3

Objectives:	$f_0(x) = -a0$	$f_1(x) = area$	
Constraints:	$g_0(x) = \frac{gbw}{35 \times 10^6} - 1$	$g_1(x) = \frac{pm}{65} - 1$	$g_2(x) = 1 - \frac{pm}{90}$

differential amplifier circuit in Fig. 4.3 using the design specifications from Table 4.1.

$$g_i(x) = \begin{cases} \frac{p_i - P_i}{|P_i|} & \text{when the constraint is } p_i \geq P_i \\ p_i & \text{when the constraint is } p_i \geq 0 \\ -p_i & \text{when the constraint is } p_i \leq 0 \\ \frac{P_i - p_i}{|P_i|} & \text{when the constraint is } p_i \leq P_i \end{cases} \qquad (4.2)$$

4.1.3 Outputs

Finally, the output is a set of feasible circuits with different sizing solutions, all fulfilling the constraints, giving the designer the possibility to choose the most appropriate tradeoff among the objectives being optimized. Figure 4.4 shows a POF obtained after the optimization of the folded cascade amplifier from Fig. 4.3, with several different sizing solutions. This set of sizing solutions allows the designer not only to explore various solutions within the solution space, choosing the one that is the most suitable, but also to save a huge time in the project execution. The points highlighted in the POF illustrate with practical values the tradeoff between objectives. For every point in the POF, the respective values for the dimensions of each device of the circuit are defined, and their layout can be generated automatically with LAYGEN II. Although, only two objectives are presented for clarity, more objectives are supported.

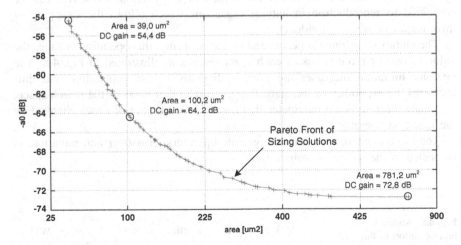

Fig. 4.4 POF obtained during the sizing task

4.2 Integration of the Gradient Model in the Optimization Kernel

The integration of the Gradient Model into GENOM-POF is done by embedding it in the crossover and mutation evolutionary operators. Each element in the population, chromosome, encodes the information of a different sizing solution, corresponding each gene to one input variable. So, each chromosome has a fixed number of genes equal to the number of optimization variables (Fig. 4.5).

As the traditional genetic algorithms (GA), the NSGA II kernel is an evolutionary optimization scheme that simulates natural evolution. It operates over a population composed by several chromosomes, each representing a different candidate solution. Each chromosome differs from each other due to their different variables' values. The genetic operators, crossover and mutation, are used to create new individuals from the initial population (usually obtained randomly), the first, by combining the genetic information from the parents, and the second by introducing random changes in the individual. The new individuals' fitness is evaluated and they are mixed with the parents and ranked. The fittest individuals are selected as the new parents, and the others discarded. The process is repeated until the ending criterion is reached (usually a fixed number of iterations). The distinguish characteristic of NSGA-II is that the ranking is made using Pareto dominance. Figure 4.6 shows the evolution of the population during one generation in a general evolutionary kernel.

The crossover operator recombines the genes in the population by generating new chromosomes combining two parent chromosomes to produce new offspring chromosomes. Figure 4.7 illustrates a common crossover operator. The idea is that the new chromosomes may correspond to better solutions than both of the parents by taking the best characteristics from each of the parents. Crossover can be compared to reproduction in natural organisms which allows the swapping of information between individuals.

The mutation operator is performed on the offspring, this operation changes the value of one or more genes of each chromosome, as illustrated in Fig. 4.8. The operator mutation increases the genetic diversity in the population, and thus increases the capability of the algorithm to search other areas of the search space. The mutation rate is the parameter that controls the number of genes changed by the mutation operator.

The implementation of the new genetic operators, crossover and mutation, is presented in the following subsections.

Fig. 4.5 Abstract representation of the chromosome in the evolutionary kernel

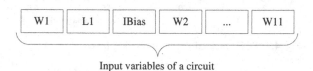

Input variables of a circuit

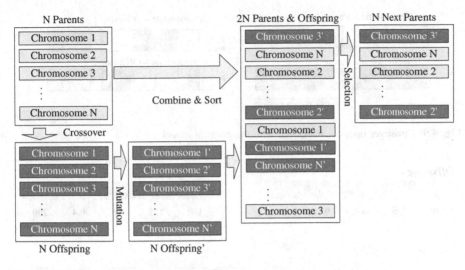

Fig. 4.6   Evolution of the population in a GA

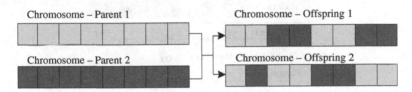

Fig. 4.7   Illustration of a crossover operator

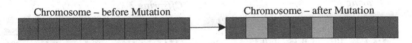

Fig. 4.8   Illustration of the mutation operator

4.2.1 Gradient Model Applied to the Crossover Operator

The first approach to integrate the Gradient Model in the optimization kernel was by applying it to the crossover operator. The gradient rules are here used to affect the genes of the parents as they are recombined. The gradient rules are used to select and change the genes, which have the desired effect on the output measures. However, the mutation operator is applied over the resultant chromosomes which can undo the changes. Figure 4.9 exemplifies an abstract integration of the Gradient Model after the crossover. The grey color represents the affected genes.

Figure 4.10 illustrates the application of a Gradient Model in the crossover operator, the selected gradient rule indicates that to increase the GBW output, both W1 and IBias should be increased. The variation applied to these variables is

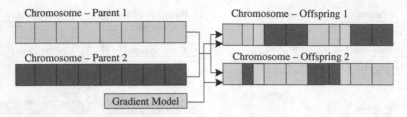

Fig. 4.9 Crossover operator integrated with the gradient model

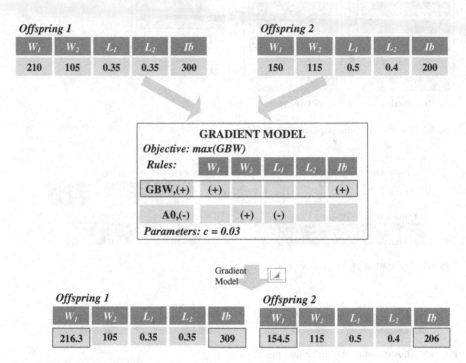

Fig. 4.10 Application example of the gradient model in the crossover

between 0 and 3 % of their value, and is defined by the Change Ratio (all changes in the illustration were done considering 3 %). The Change Ratio is one of the model parameter, the other is the Apply Rate which controls the application or not of the model during crossover, later, in the subsection Graphical User Interface, those parameters will be explained in detail.

4.2.2 Gradient Model Applied to the Mutation Operator

Another approach to integrate the Gradient Model in the optimization kernel is to integrate the model into the mutation operator. This approach do not fall into the same problem of applying the model to the crossover operator, which can drive chromosomes for areas of optimization and then that direction be changed by the intervention of the mutation operator.

As seen before, the chromosome is represented by the vector of continuous variables $\{x_1,\ldots, x_n\}$ representing the design variables. To speed up the convergence of the algorithm, the knowledge extracted from the gradient model is used to make the mutation operator more efficient. The reference mutation operator in GENOM-POF uses the continuous valued operator introduced by Deb and Goyal in [7]. In this mutation operator, δ_i is the mutation perturbation applied and is defined as $\delta_i = \left(x_i^M - x_i\right)/\left(X_i^{Max} - X_i^{Min}\right)$, where x_i^M and x_i are the mutated and original values, respectively. Moreover, δ_i is a random variable, with values in the interval of -1 to 1, and probability density function given by (4.3), where, η is a parameter used to control the distribution's spread. Figure 4.11 shows the $p.d.f.$ for various values of η.

$$P(\delta) = 0.5 \times (\eta + 1) \times (1 - |\delta|)^{\eta} \tag{4.3}$$

A factor of disturbance $\bar{\delta}$ of δ can be obtained from a uniform random number $u \in [0, 1]$ using (4.3), which is obtained from (4.4) by solving $\int_{-1}^{\bar{\delta}} P(\delta) = u$.

$$\bar{\delta} = \begin{cases} (2u)^{\frac{2}{\eta+2}} - 1, & \text{if } u < 0.5 \\ 1 - [2(1-u)]^{\frac{2}{\eta+2}}, & \text{if } u \geq 0.5 \end{cases} \tag{4.4}$$

Then, the mutated value, x_i^M, is given by $x_i^M = x_i + \bar{\delta}(X_i^{Max} - X_i^{Min})$. The gradient rules obtained during the generation of the gradient model are then applied. The application of the rules follows (4.5), where x_i^G is the variable value after the application of the rule, $\gamma(Si_i)$ is a function of the gradient symbol defined in (4.6), and $u \in [0, c]$ is a uniformly distributed random number between 0 and c, the Change Rate model parameter.

$$x_i^G = (1 + \mu.\gamma(Si_i)) x_i^M \tag{4.5}$$

Fig. 4.11 Probability distribution for creating a mutated value [7]

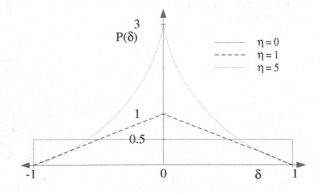

Individual: after mutation

W_1	W_2	L_1	L_2	Ib
210	105	0.35	0.35	300

Gradient Model

Individual: after gradient model

W_1	W_2	L_1	L_2	Ib
216.3	105	0.35	0.35	309

GRADIENT MODEL

Objective: max(GBW)

Rules:

	W_1	W_2	L_1	L_2	Ib
GBW,(+)	(+)				(+)
A0,(-)		(+)	(-)		

Parameters: c = 0.03

Fig. 4.12 Example of application of the gradient model in the mutation

$$\gamma(Si_i) = \begin{cases} +1 \; when \; Si_i = (+) \\ -1 \; when \; Si_i = (-) \end{cases} \qquad (4.6)$$

The application of the gradient rules is dependent on the existence of a suitable rule for the optimization targets, i.e., it is irrelevant to have a rule to decrease the gain, if the optimization target is to increase it. The rules are selected by searching if there is a rule that causes the desired effect on each optimization objective. If this rule is found, then the variables with larger contributions are affected as described before. Figure 4.12 shows an example of applying a gradient rule.

Once again, in the example the Gradient Model indicates to increase the values of W1 and IBias, and with a maximum percentage of changing the values of 3 %. After the model application the chromosome may have incorporated the knowledge needed to achieve faster the optimal solution. With the introduction of circuit knowledge through the gradient it is expectable to reach sub-optimal solutions faster than GENOM-POF without model, or even to reach optimal solutions that GENOM-POF cannot reach alone. Figure 4.13 presents an example of the mutation operator integrated with Gradient Model.

Fig. 4.13 Solutions search space for mutation integrated with gradient model

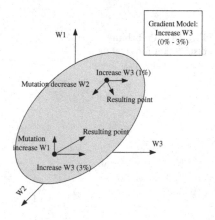

4.3 Graphical User Interface (GUI)

The AIDA environment integrated with the Gradient Model, framework and GUI are implemented in Java^{TM} 1.6. This GUI provides a simple and fast way for the designer to set the values of constraints and the range of the optimization variables, also through the GUI the designer can set the type of optimization to be made on objectives (maximize or minimize), and monitor the results at any stage of the evolutionary generation.

Figure 4.14 presents the overview of the AIDA's GUI. From this overview is already possible to identify in the upper left the schematic of the circuit in use. Beneath the schematic are the options available to the user for controlling the circuit, model and algorithm, which will be presented in detail below. In the right, on top the estimated layout and on bottom the details, of each of the solutions in the POF listed in the center, are presented.

Figure 4.15 shows possible objectives (DC Gain and area) for a design, how-ever, these can be chosen by the designer through the GUI since they are defined in the netlist. Also, using the GUI the designer can define the type of objective intended to maximize or minimize.

In Fig. 4.16 it is possible to see the constraints of design and their limits. The limits and type of the constraints ($>=$ or $<=$) can be easily changed. Also, the output being limited can be chosen from any of the available performance measures.

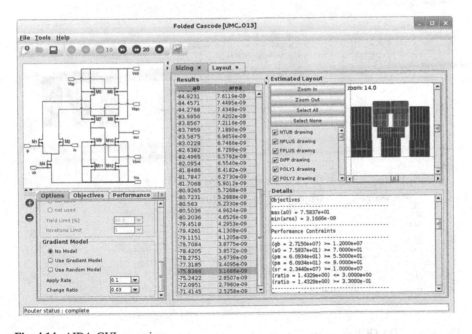

Fig. 4.14 AIDA GUI: overview

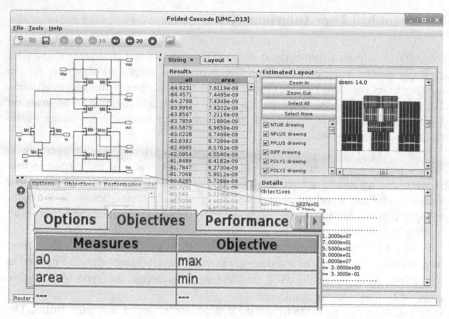

Fig. 4.15 AIDA GUI: objectives

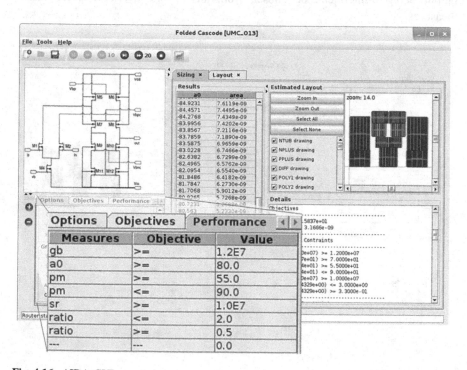

Fig. 4.16 AIDA GUI: constraints

These two panels allow the designer to make several studies for the same circuit, by changing the values of the objectives and their constraints. However, the designer can also change the search space for the optimization problem by changing the range of the input variables. Figure 4.17 shows the several optimization variables considered in optimization of this circuit, and their respective minimum and maximum range. The designer can change these ranges values to his own desire by introducing those values in the GUI. Changing the ranges of the

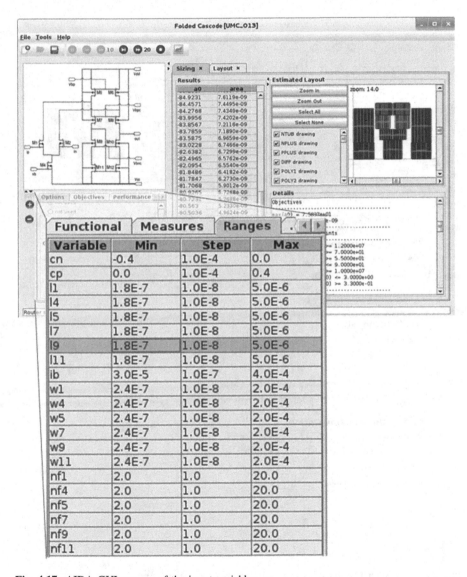

Functional	Measures	Ranges		
Variable	**Min**	**Step**	**Max**	
cn	-0.4	1.0E-4	0.0	
cp	0.0	1.0E-4	0.4	
l1	1.8E-7	1.0E-8	5.0E-6	
l4	1.8E-7	1.0E-8	5.0E-6	
l5	1.8E-7	1.0E-8	5.0E-6	
l7	1.8E-7	1.0E-8	5.0E-6	
l9	1.8E-7	1.0E-8	5.0E-6	
l11	1.8E-7	1.0E-8	5.0E-6	
ib	3.0E-5	1.0E-7	4.0E-4	
w1	2.4E-7	1.0E-8	2.0E-4	
w4	2.4E-7	1.0E-8	2.0E-4	
w5	2.4E-7	1.0E-8	2.0E-4	
w7	2.4E-7	1.0E-8	2.0E-4	
w9	2.4E-7	1.0E-8	2.0E-4	
w11	2.4E-7	1.0E-8	2.0E-4	
nf1	2.0	1.0	20.0	
nf4	2.0	1.0	20.0	
nf5	2.0	1.0	20.0	
nf7	2.0	1.0	20.0	
nf9	2.0	1.0	20.0	
nf11	2.0	1.0	20.0	

Fig. 4.17 AIDA GUI: ranges of the input variables

design variables changes the search space, it reduces when the ranges are
decreased and expands as the ranges increase. The point to keep in mind is that
with narrower search space, the algorithm converges faster to the optimal solutions, while with a larger search space the same algorithm will have more difficulty
in finding the optimal solutions.

Other options like the type of strategy to use or the possibility to use or not the
Gradient Model are included in GUI, also the designer can setup the parameters of
the NSGA-II and the Gradient Model. Figure 4.18 shows the type of strategy to be

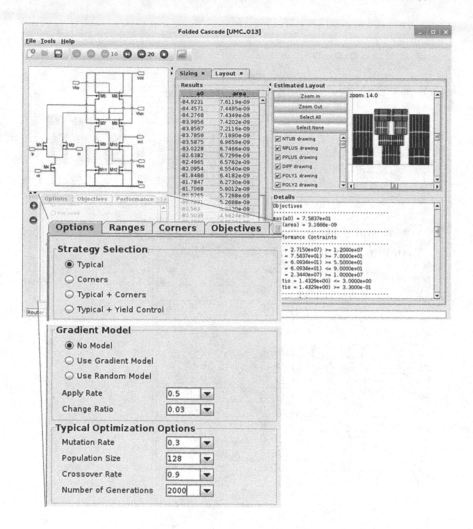

Fig. 4.18 AIDA GUI: gradient model and optimization options

adopted for the optimization problem, GENOM-POF supports both typical and corner optimization, however, in this work only the typical strategy is considered.

The following settings, in Fig. 4.18, are: the use or not of the Gradient Model or the Random Model; the model application parameters, respectively, Apply Rate, which is the frequency/percentage of use of the model chosen in the optimizer, and Change Ratio, which is the maximum percentage of change of the value of a variable. The Random Model will be presented in the next chapter, together with the results, and is used as reference for comparison with the Gradient Model; the last options are the NSGA-II parameters, the mutation rate controls the percentage of genes that are mutated in each generation and influences the diversity of the population, the population size is the number of individuals (chromosomes) in a population participating in the evolutionary process, and the crossover rate defines the frequency/percentage of the crossover operation, and, finally, the number of generations, which corresponds to the number of execution cycles of the algorithm.

4.4 Conclusions

This chapter introduced the architecture of the AIDA environment for the automatic synthesis of analog integrated circuits. The more specific architecture of GENOM-POF was also presented, which is the tool for circuit sizing optimization. All the specifications like inputs, structure of the optimization kernel, design strategies, and the outputs were explained. The integration of the model into the optimization kernel was detailed described. The focus was kept on the evolutionary operators since the Gradient Model is applied through the crossover or mutation operator. Finally, this chapter also introduced the AIDA's GUI.

References

1. N. Lourenço, N. Horta, GENOM-POF: multi-objective evolutionary synthesis of analog ICs with corners validation, in *Proceedings of the fourteenth international conference on Genetic and evolutionary computation conference (GECCO '12)*, pp. 1119–1126. July 2012. doi:10.1145/2330163.2330318
2. R. Martins, N. Lourenco, S. Rodrigues, J. Guilherme, N. Horta, AIDA: Automated analog IC design flow from circuit level to layout, in *Synthesis, Modeling, Analysis and Simulation Methods and Applications to Circuit Design (SMACD), International Conference on*, pp. 29–32, Sept 2012. doi:10.1109/SMACD.2012.6339409
3. R. Martins, N. Lourenço, N. Horta, *Generating Analog IC Layouts with LAYGEN II*, SpringerBriefs in Applied Sciences and Technology–Computational Intelligence, (Springer, New York, 2013)
4. R. Martins, N. Lourenço, N. Horta, LAYGEN II-automatic layout generation of analog integrated circuits, for publication in *IEEE Transactions on Computer-Aided Design of Integrated Circuits and Systems*, doi: 10.1109/TCAD.2013.2269050

5. R. Martins, N. Lourenço, N. Horta, Routing analog ICs using a multi-objective multi-constraint evolutionary approach. Analog Integr. Circuits Signal Process. 1–13 (2013). doi:10.1007/s10470-013-0088-9
6. K. Deb, A. Pratap, S. Agarwal, T. Meyarivan, A fast and elitist multiobjective genetic algorithm: NSGA-II. IEEE Trans. Evol. Comput. **6**(2), 182–197 (2002)
7. K. Deb, M. Goyal, A combined genetic adaptive search (GeneAS) for engineering design. Comput. Sci. Inform. **26**(4), 30–45 (1996)

Chapter 5
Results

Abstract This chapter illustrates the application of the proposed methodology to practical examples. The framework of the proposed methodology for the automatic generation of analog integrated circuits (IC) layout has been coded in JAVA and was executed, for the presented examples, on an Intel® Core™ 2 Quad CPU 2.4 GHz with 6 GB of RAM.

Keywords Analog IC design · Circuit-level sizing · Evolutionary computation · Gradient model

5.1 POFs Analysis

In this study, two objectives are considered to illustrate the multi-objective nature of the proposed optimization approach, which also facilitates a clear graphical representation of the Pareto optimal fronts (POF). In order to compare GENOM-POF [1] with the integrated solution GENOM-POFGM (GENOM-POF + Gradient Model) the following POF performance indicators are defined:

- Non-dominated area of the POF;
- Number of points in the POF;
- Standard deviation of points on the $f_1(x)$ axis of the POF;
- Standard deviation of points on the $f_2(x)$ axis of the POF.

Before entering into the details of the performance indicators, it is important to grasp the meaning of Pareto dominance, illustrated in Fig. 5.1. The definition of dominance states that a solution x_i dominates a solution x_j, if both the following conditions are true:

F. A. E. Rocha et al., *Electronic Design Automation of Analog ICs Combining Gradient Models with Multi-Objective Evolutionary Algorithms*, SpringerBriefs in Computational Intelligence, DOI: 10.1007/978-3-319-02189-8_5, © The Author(s) 2014

Fig. 5.1 Pareto dominance: x_2 is dominated by x_1 while x_1 and x_3 are not dominated, assuming both f_1 and f_2 are being minimized

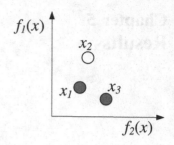

- The solution x_i is not worse than x_j in any objectives;
- The solution x_i is strictly better than x_j in at least one objective.

If any of the above conditions is violated the solution x_i does not dominate the solution x_3. For a given set of solutions, all possible pairwise comparison should be performed in order to find which of those are the non-dominated solutions forming the Pareto front.

Either the dominated or the non-dominated area of the POF are extremely relevant indicators to make a quantitative comparison among POFs and are computed as follow:

- The axis are normalized to be between 0 and 1;
- The dominated area, illustrated in Fig. 5.2 by A, is calculated by the sum of trapezoidal areas defined by successive POF elements as described by expression (5.1):

$$A = \sum \left(\frac{(1-a)+(1-b)}{2} \right) \times h \tag{5.1}$$

- The non-dominated area, illustrated in Fig. 5.2 as area B, was defined as being $B = 1 - A$. Thus, the smaller the non-dominated area, the better is the generated POF.

Fig. 5.2 Dominated and non-dominated areas

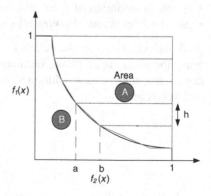

5.2 Circuit Under Test: Single-Ended Folded Cascode Amplifier

The single-ended folded cascade amplifier, which schematic and test bench are presented in Fig. 5.3, is here used to compare GENOM-POF with GENOM-POFGM. In all the studies, the μ UMC (United Microelectronics Corporation Group) complementary metal–oxide–semiconductor (CMOS) technology was considered. Two different studies were performed, the first considering 15 input variables (15th dimensional search space), which corresponds to a large solution search space, and the second reducing the input variables to 12 (12th dimensional).

Fig. 5.3 Schematic and testbench of the single-ended folded cascade amplifier

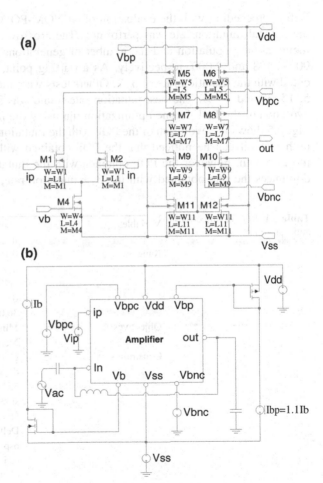

5.3 Case Study I: 15 Input Variables

The first case study was performed considering 15 input/optimization variables, 2 objectives and the performance and functional constraints, as defined in Table 5.1. The optimization variables include the *W's* and *L's* of all transistors, the cascode bias tensions, represented by *vbcp* and *vbcn* respectively, and the bias current. This optimization task is executed by GENOM-POF and by GENOM-POFGM for the very same conditions.

5.3.1 GENOM-POF

Before proceeding with the evaluation of GENOM-POFGM, a brief study to tune the optimal mutation rate was performed. The crossover rate, the number of elements in the population and the number of generations were fixed to the values 90 %, 128 and 2,000, respectively. As a starting point, GENOM-POF was executed with the mutation rate of 3 %. Others tests were made for a mutation rate of 5, 15, 30 and 45 %. The best mutation rate found was 30 %, Fig. 5.4 shows the evolution of the POF in one optimization run using the mutation rate of 3 %, and Fig. 5.5 shows the evolution of the POF with the mutation rate of 30 %. Observing both plots, it can be noticed that the POF obtained with a 3 % mutation rate is more smooth, however the POF obtained with the mutation rate of 30 % clearly dominates the one obtained with the 3 % mutation rate, for the same number of

Table 5.1 Range, objectives and constraints		
	Variables	vbcn, vbcp, l1, l4, l5, l7, l9, l11, ib, w1, w4, w5, w7, w9, w11
	Ranges	$0.18~\mu m <= l^* <= 5.0~\mu m$
		$0.24~\mu m <= w^* <= 200.0~\mu m$
		$-400~mV <= vbcn <= 0.0~V$
		$0.0~V <= vbcp <= 400~mV$
		$30.0~\mu A <= ib <= 400.0~\mu A$
	Objectives	Min(area)
		Max(a0)
	Constraints	gb >= 12 MHz
		dc_gain >= 80 dB
		$55° <= pm <= 90°$
		sr >= 10 V/μs
		Overdrive_m(*) >= 30 mV
		Delta_m(*) >= 1.2
		osp >= 300 mV
		osn <= −300 mV
		(*) the constraints apply to: M1, M4, M5, M7, M9 and M11

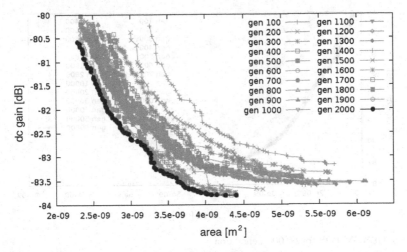

Fig. 5.4 GENOM-POF with mutation rate: 3 % and for 2,000 generations

generations. This difference is due to the fact that smaller mutation rates lead to early convergence issues; moreover, the mutation operator used, unlike other approaches where the new genes are set as random value within the allowed range as the new genes, generates new genes that are in average close the original values, thus the better performance when using a higher mutation rate.

With the mutation rate tuned, an exhaustive optimization with 60,000 generations was performed; the obtained POF is shown in Fig. 5.6. Both POFs provide a reference to the evaluation of the performance of GENOM-POFGM.

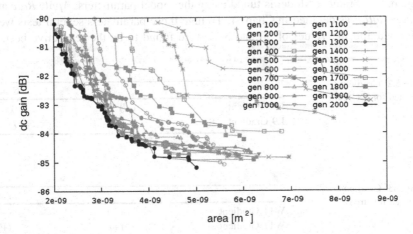

Fig. 5.5 GENOM-POF with mutation rate: 30 % and for 2,000 generations

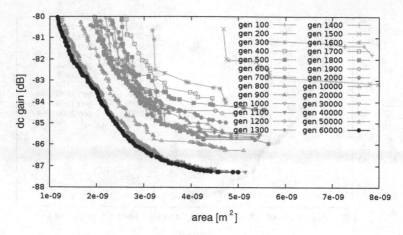

Fig. 5.6 GENOM-POF for 60,000 generations

5.3.2 GENOM-POFGM

For this example a Gradient Model was generated considering a Design of Experiments (DOE) matrix with base two (B = 2) and the contribution of one optimization variable (N = 1). The resulted Gradient Model for both objectives is shown in Tables 5.2 and 5.3.

The generated Gradient Model shows a positive gradient for both objectives, i.e., for the maximization of the DC gain the model tells us to increase the value of the variable *L9,* and for the minimization of area the model indicates to the algorithm to decrease the value of *W11.* The generated model for this example took less than 5 min to be generated and the model can be reused for the same circuit and optimization variables but with different values of constraints. The usage of the Gradient Model is tuned using the model parameters: *Apply Rate* and *Change Ratio,* as detailed in Chap. 4. To tune these parameters several tests were performed and the values of 50 and 3 % were found to be the respective best.

Table 5.2 Gradient generated for DC gain			DC gain
L9 Gradient		(−)	(−)
L9 Gradient		(+)	(+)

Table 5.3 Gradient generated for area			Area
W11 Gradient		(−)	(−)
W11 Gradient		(+)	(+)

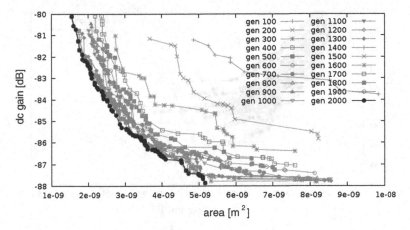

Fig. 5.7 GENOM-POFGM (apply rate $= 50$ % and change ratio $= 3$ %) for 2,000 generations

GENOM-POFGM was then used to optimize the same circuit in the same conditions used for GENOM-POF (mutation rate of 30 %, crossover rate of 90 % and population size of 128 and 2,000 generation). The POF obtained is shown in Fig. 5.7.

Figure 5.8 shows the POFs obtained from 2,000, 4,000 and 60,000 optimization generations in GENOM-POF, and, superimposed, the POF obtained from 2,000 optimization generations in GENOM-POFGM. The POF obtained with GENOM-POFGM, with only 2,000 generations, clearly dominates the ones obtained with GENOM-POF for 2,000 and 4,000 generations. Even the POF obtained after an exhaustive optimization with 60,000 generations, thought generically better, does not dominate completely the POF obtained with GENOM-POFGM, not reaching the maximum value for DC Gain reached by GENOM-POFGM. These results

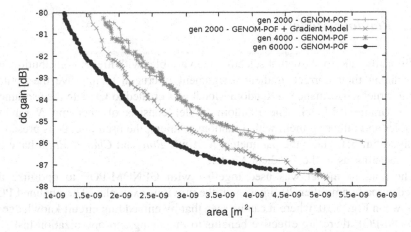

Fig. 5.8 GENOM-POF (60,000, 4,000, and 2,000 gen.) versus GENOM-POFGM (2,000 gen.)

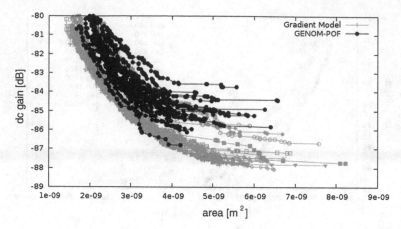

Fig. 5.9 20 different initial populations for comparison between GENOM-POF and gradient model

show the effectiveness of the knowledge captures by the Gradient Model to achieve better solutions faster than the GENOM-POF for the same number of generations.

The 2,000 generations were executed in approximately 30 min, for the 4,000 generations the time doubles, and for the 60,000 generations the optimization process takes approximately 15 h. GENOM-POFGM for 2,000 generations shows competitive results in comparison with GENOM-POF for 60,000 generations in approximately less 14 h and 30 min. To confirm that this is not an isolated case, 20 executions with different seeds were performed. The results are shown in Fig. 5.9, where it can be seen that the inclusion of the gradient model consistently lead to better solutions.

5.3.3 Random Model

Additionally, and to show that selecting the variables with higher contribution and determining their correct gradient assignment is crucial to improve the optimization kernel performance, a Random Model was created to validate the usefulness of the Gradient Model. The Random Model consists of choosing N random variables and affecting them with a random gradient, the pseudo-code is presented in Algorithm 5.1 where the parameters N, *Apply Rate* and *Change Ratio* have the same meaning as in the Gradient Model.

The random model was used together with GENOM-POF to optimize the folded cascade amplifier 20 times as in the previous examples. The obtained POF is shown in Fig. 5.10, where it can be seen that by embedding circuit knowledge in GENOM-POF there are effective benefits to the sizing and optimization task.

Algorithm 5.1 Random model

```
1.      Set the parameters B, N
2.      for each output yⱼ : j=1,…,M do
2.1       if apply_rate > random(0, 100) then
2.2         Cⱼ = selectRandomVars(N)
2.3         for each input xᵢ in Cⱼ : i=1,…,N do
2.3.1         if random(0,1) == 0 then
2.3.2           Δ = (100 + random(0, change_ratio))/100
2.3.3         else
2.3.4           Δ = (100 - random(0, change_ratio))/100
2.3.5         xᵢ = Δ.xᵢ
```

5.3.4 Comparison of Different Optimization/Sizing Approaches

To quantify the graphical analysis described this far, Tables 5.4, 5.5 and 5.6 summarize for each run the following numerical and statistical indicators:

- Low Area: Finding the minimum circuit area is one of the objectives of this design problem, so the lowest area reached is an indicator that a designer would take into consideration. This indicator is represented by the coordinates in the objective space (*area, dc gain*) of the solution with the lowest area.
- Max DC Gain: Finding the maximum DC gain is another objective of this design problem, and is also used an indicators represented by the coordinates in the objective space (*area, dc gain*) of the solutions with the greatest DC gain.

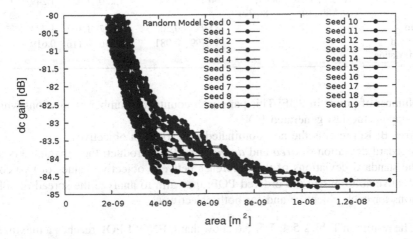

Fig. 5.10 Random model for 20 different initial populations

Table 5.4 POFs (20 different seeds) analysis for GENOM-POF

Population: 128 Mutation: 30 % Crossover: 90 % Nr. of Generations 2,000

Run ID	Low area [μm^2, dB]	Max DC gain [μm^2, dB]	# Points	Area: $1 - A = B$	σ_{area}	$\sigma_{dc\ gain}$	$\sigma_{area} \times \sigma_{dc\ gain}$
0	(2008, 80.21)	(5004, 85.19)	63	0.435	1.523	1.278	1.947
2	(**1676**, 80.67)	(4528, 85.36)	48	0.390	1.383	1.185	1.640
...	...	...	...	...	...	...	...
12	(1736, 80.56)	(4204, **86.81**)	47	0.231	1.302	0.837	1.091
19	(2304, 80.76)	(4963, 84.48)	46	0.511	1.093	1.118	1.223
Mean			51.55	0.426	1.549	1.188	1.842
Standard deviation			8.438	0.085	0.479	0.163	0.621

Table 5.5 POFs (20 different seeds) analysis for gradient model

Population: 128 Mutation: 30 % Crossover: 90 % Nr. of Generations 2,000

Run ID	Low area [μm^2, dB]	Max DC gain [μm^2, dB]	# Points	Area: $1 - A = B$	σ_{area}	$\sigma_{dc\ gain}$	$\sigma_{area} \times \sigma_{dc\ gain}$
0	(1544, 80.10)	(5167, 87.85)	85	0.134	1.346	1.172	1.578
3	(**1426**, 80.18)	(6881, 87.20)	87	0.197	1.732	1.168	2.024
...	...	...	...	...	...	...	...
11	(1596, 80.05)	(6493, **88.07**)	95	0.117	1.197	1.159	1.388
19	(2257, 80.26)	(5647, 86.30)	66	0.348	1.220	1.023	1.249
Mean			81.7	0.200	1.527	1.162	1.784
Standard deviation			16.799	0.081	0.496	0.116	0.619

- Number of points in POF: This parameter counts the number of solutions which compose the last generated POF.
- Area B: Represents the non-dominated area for both objectives.
- Standard deviation of *area* and *dc gain* and their product: these parameters are the standard deviations of the difference between objective values of two consecutive solutions in the ordered POF; they aim to analyze the spread of solutions for each objective and for both objectives.

The results in Tables 5.4, 5.5, 5.6 show that GENOM-POF reaches a maximum DC Gain around 86.81 dB for the seed 12 and a minimum area around 1676 μm^2

Table 5.6 POFs (20 different seeds) analysis for random model

Population: 128 Mutation: 30 % Crossover: 90 % Nr. of Generations 2,000

Run ID	Low area [μm^2, dB]	Max DC gain [μm^2, dB]	# Points	Area: 1 − A = B	σ_{area}	$\sigma_{dc\ gain}$	$\sigma_{area} \times \sigma_{dc\ gain}$
0	(2731, 80.18)	(7013, 84.29)	98	0.482	1.604	1.093	1.755
7	(**1846**, 80.11)	(4031, 84.42)	117	0.407	1.185	1.388	1.645
...	...	...	...	...	...	...	...
10	(2102, 80.33)	(11480, **84.86)**	91	0.416	2.639	1.196	3.1589
19	(2.359, 80.51)	(8147, 84.20)	72	0.467	3.592	1.260	4.529
Mean			93.8	0.455	2.444	1.281	3.156
Standard deviation			16.516	0.029	0.873	0.155	1.253

in the seed two. GENOM-POFGM reaches the maximum DC Gain of 88.07 dB at seed 11 and the minimum area 1430 μm^2 in the seed three. Finally, the GENOM-POF plus the Random Model achieves the maximum of 84.86 dB at seed 10 and the minimum area 1846 μm^2 in the seed seven.

However, all these maximum and minimum values are achieved in different seeds. So, the mean of the non-dominated areas becomes a relevant value to different approaches. By observing these tables, the Gradient Model presents very good results in terms of non-dominated area by having the lowest areas.

Finally, the observation of the non-dominated area, the number of points in the POF and the standard deviations leads to the conclusion that GENOM-POFGM performs significant better than the other tested approaches. Table 5.7 summarizes all the comparisons between the gradient and random models and the reference GENOM-POF implementation.

5.4 Case Study II: 12 Input Variables

Using the same circuit as before, but now with a number of optimization variables reduced to 12, by removing the biasing variables ib, vbcn and vbcp, while keeping the objectives and constraints.

Table 5.7 Comparison between the gradient and random models and GENOM-POF

	Non-dominated area	Nr. points POF	δ_{area}	$\delta_{dc\ gain}$	$\delta_{area} \times \delta_{dc\ gain}$
Gradient model	Better than GENOM-POF	Better than GENOM-POF	Similar to GENOM-POF	Similar to GENOM-POF	Similar to GENOM-POF
Random model	Worse than GENOM-POF	Better than GENOM-POF	Worse than GENOM-POF	Similar to GENOM-POF	Worse than GENOM-POF

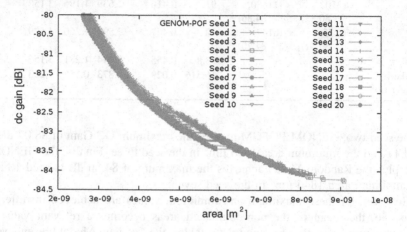

Fig. 5.11 GENOM-POF optimization for case study II

5.4.1 GENOM-POF

Figure 5.11 illustrates the GENOM-POF optimization runs with the mutation rate of 30 %, crossover rate of 90 %, population size of 128, the number of generations of 2,000 for 20 different runs.

It is clear that results are now below the ones in the previous example (shown in Fig. 5.6). This is due to the reduction in the number of optimization variables, especially the fixed biasing of the circuit. The previous example showed a maximum DC Gain around 85 dB and a minimum area around 2000 μm^2 however, in this example the maximum DC Gain is around the 84 dB and the minimum area is around 3000 μm^2.

			DC gain
Table 5.8 Gradient rules generated for DC gain	L11 Gradient	(−)	(−)
	L11 Gradient	(+)	(+)

			Area
Table 5.9 Gradient rules generated for area	W5 Gradient	(−)	(−)
	W5 Gradient	(+)	(+)

5.4.2 GENOM-POFGM

The Gradient Model for this example was generated like the one generated in the previous example, and is shown in Tables 5.8 and 5.9.

Figure 5.12 shows the 20 POFs obtained for 20 different optimization runs with GENOM-POFGM. Like the previous example the Apply Rate is 50 % and the Change Ratio is 3 %.

Like the result obtained with GENOM-POF, this result, in comparison with the results shown in Fig. 5.8, presents a real deterioration of the solutions. For this optimization the maximum DC Gain obtained is around the 84 dB while before was around 88 dB. Also for the area measure the solutions found are worse, around the 3000 μm^2 nm while before was around 2000 μm^2.

The results for both GENOM-POF and GENOM-POFGM are worse in this case than in the case with 15 optimization variables. This deterioration is explained with the fixed variables ib, *vbcn* and *vbcp*, which limits the range of operation of the circuit.

5.4.3 Comparison of Different Optimization/Sizing Approaches

The first way to compare GENOM-POF and GENOM-POFGM is performed by analyzing their POFs. Figure 5.13 shows the overlay of both POFs in the same plot. It does not show significant improvements, as both results seem very similar.

Besides the visual analysis of the POFs does not allow any conclusion about the improvement or not with the Gradient Model, the statistical study does confirm the contribution of the proposed approach. Tables 5.10 and 5.11 show the analyses performed for GENOM-POF and GENOM-POFGM respectively. As observed before in the POFs, Tables 5.10 and 5.11 do not show significant improvements by the integration of the Gradient Model. The lowest value of area of 2607 $\mu\mu m^2$ reached by GENOM-POF and 2577 μm^2 reached by GENOM-POFGM, are, in practical terms, the same. The same happens to the maximum DC Gain, GENOM-

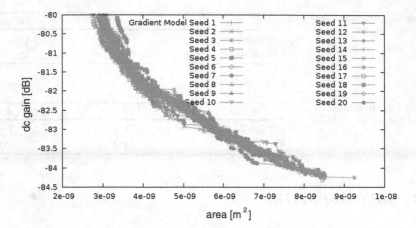

Fig. 5.12 GENOM-POFGM optimization for case study II

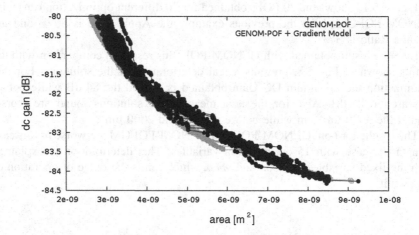

Fig. 5.13 Comparison between GENOM-POF and GENOM-POFGM

POF presents the maximum of 84.1 dB while GENOM-POFGM achieves the maximum DC Gain of 84.41 dB.

However, a careful analysis of the non-dominated area (defined as area B) show that the non-dominated area is generically lower for the case of GENOM-POFGM. To further study this phenomenon, for each seed, the POFs from the generation 500 to the generation 2,000 were analyzed showing that GENOM-POFGM had consistently less non-dominated area, which means that reaches better solutions faster than GENOM-POF. The summary of these results is presented in Table 5.12.

Table 5.10 POFs (20 different seeds) analyses for GENOM-POF

Population: 128 Mutation: 30 % Crossover: 90 % Nr. of Generations 2,000

Run ID	Low area [μm^2, dB]	Max DC gain [μm^2, dB]	# Points	Area: $1 - A$ $= B$	δ_{area}	δ_{dc} gain	$\delta_{area} \times \delta_{dc\ gain}$
0	(2799, 80.00)	(8345, **84.10**)	129	0.287	1.297	0.638	0.828
...	...	...	...	...	...	...	...
11	(**2607**, 80.02)	(7501, 83.92)	130	0.293	1.593	0.605	0.965
19	(2646, 80.04)	(6590, 83.56)	128	0.333	1.640	0.658	1.079
Mean			128.75	0.329	1.467	0.659	0.970
Standard deviation			1.2085	0.046	0.374	0.054	0.264

Table 5.11 POFs (20 different seeds) analyses for gradient model

Population: 128 Mutation: 30 % Crossover: 90 % Nr. of Generations 2,000

Run ID	Low Area [μm^2, dB]	Max DC Gain [μm^2, dB]	# Points	Area: $1 - A$ $= B$	δ_{area}	δ_{dc} gain	$\delta_{area} \times \delta_{dc\ gain}$
0	(2647, 80.01)	(7238, 83.77)	128	0.308	1.789	0.729	1.305
2	(**2577**, 80.02)	(8784, 84.41)	128	0.241	1.583	0.533	0.843
...	...	...	...	...	...	...	...
19	(2755, 80.25)	(8049, 84.23)	128	0.262	1.718	0.693	1.191
Mean			128.4	0.274	1.629	0.675	1.106
Standard deviation			0.882	0.029	0.206	0.093	0.239

Table 5.12 Analyze of non-dominated area

	Nr. of time that have less non-dominated area then GENOM-POF	Percentage that have less non-dominated area then GENOM-POF	Total number of POFs analyzed
Gradient model	270	84.64 %	319

Table 5.13 Comparison between GENOM-POF and GENOM-POFGM

	Advantages	Disadvantages
GENOM-POF	Good application to a unknown circuit Good ability to adapt to any problem Expandable to n dimensional space	The execution time can be high, because the entire analysis requires many evaluations of the outputs during the execution of internal GA The user has no control over the optimization
GENOM-POFGM	Time model generation greatly reduced (even negligible), for both simple and complex circuits Simple and functional implementation Possibility for the designer to change the gradient of the variables, the rate of application of the model and the rate of change the value of the variables	Offers robustness for problems where the search space is large

5.5 Conclusions

In this chapter the proposed approach was tested with different scenarios showing its effective contribution to improve results and accelerate convergence.

First, an example considering an optimization for 15 input variables was performed, which corresponds to a wide search space of solutions. For this example the performance of GENOM-POF, GENOM-POFGM and GENOM-POF integrated with Random Model was analyzed. The conclusions were quite clear and proved that GENOM-POFGM presents significant improvements for the sizing/ optimization task. This study also shows that GENOM-POFGM reaches a better solution for the maximization of DC Gain. In a second example, the number of variables to be optimized was reduced to 12, and the reduction in the problem complexity also lead to less significant improvement with GENOM-POFGM. In Table 5.13 some closing remarks are presented.

Reference

1. N. Lourenço, N. Horta, in *GENOM-POF: Multi-Objective Evolutionary Synthesis of Analog ICs with Corners Validation*. GECCO' 12: Proceedings of the 14th International Conference on Genetic and Evolutionary Computation Conference, July 2012, pp. 1119–1126

Chapter 6
Conclusions and Future Work

Abstract The proposed methodology for the enhancement of a state-of-the-art circuit-level synthesis approach, GENOM-POF [1], by incorporating a gradient model into a multi-objective multi-constraint optimization kernel was proved by the implementation of a tool, GENOM-POFGM (GENOM-POF + Gradient Model), which is able to generate robust circuit sizing solutions. This chapter presents the closing remarks, and the future directions for the continuous development of GENOM-POFGM.

Keywords Analog IC design · Circuit-level sizing · Electronic design automation · Computer-aided-design

6.1 Conclusions

The presented methodology corresponds to an innovative integrated circuit (IC) design automation approach by embedding a simple but effective design knowledge model, Gradient Model, into the evolutionary optimization kernel of a state-of-the-art analog circuit-level sizing tool. The new technique proved to be capable to accelerate and reduce the execution time of the circuit-level optimizer GENOM-POF. This integration of the Gradient Model with GENOM-POF enhances the optimizer efficiency, forwarding the data to the desired objectives and causing a significant reduction in the number of electrical simulations, i.e., required evaluations.

The model generation was performed using a Design of Experiments sampling technique, with two alternative strategies, Full Factorial Design and Fractional Factorial Design, and both showed no contradictions in their statistical analysis.

The Gradient Model has as main goal the description of a set of simple gradient rules, providing the designer with a direct analysis showing the contribution of the input variables to the desired objectives and/or performance measures. The model also offers a set of parameters which the user can explore and vary to adapt to the

F. A. E. Rocha et al., *Electronic Design Automation of Analog ICs Combining Gradient Models with Multi-Objective Evolutionary Algorithms*, SpringerBriefs in Computational Intelligence, DOI: 10.1007/978-3-319-02189-8_6, © The Author(s) 2014

proposed problem, using the graphical user interface. This optimizer represents a totally automated alternative to the traditional optimization techniques, where the execution time is usually extremely high.

The Gradient Model integrated with the mutation operator of the genetic algorithm proved to be useful to bias the search direction into the most promising direction. The model application has been proved through the presentation of a complex case study. This case study was divided in two different problems, where in the first exists a large solution space and in the second the solutions space is reduced. These two examples validated the fact that Gradient Model integrated in GENOM-POF presents better solutions for large solutions space; and also, these examples proved that even for the worst case, small solution space, the Gradient Model does not worsened the results of GENOM-POF and still get small improvements over the GENOM-POF results.

Finally, the proposed objectives for this work were achieved and a new optimizer was created.

6.2 Future Work

In analog IC design automation, there is still a long way to end in this domain; the improvement on productivity of analog design is a demand of economic market. Based on this work and its large application on analog design, there are some suggestions for future research which may improve even more its efficiency.

The first suggestion is the application of Gradient Model for the Corners validation. The second suggestion is to improve the accuracy of the model by performing an extra sample step of the circuit after reaching the first Pareto optimal front (POF). Several other opportunities can easily be pointed out for future work showing the large potential of the presented approach.

The integration of the model in the GENOM-POF optimization kernel can also be performed in alternative ways. An alternative is its application to only one objective variable, other alternative approach is an application of the Gradient Model that is not always the same. Figure 6.1 shows an approach where the model is applied to just one of the objectives half of the time (25 % each of the objectives), while the other half of the time it is applied to all the objectives. The expected result with this alternative approach is to accelerate the optimization process by the application of the model to all the optimization objectives, and at the same time to maximize/minimize even more the objectives by the single application of the model to a specific objective.

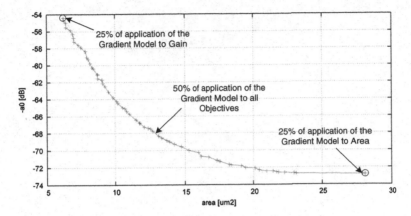

Fig. 6.1 Alternative approach to apply the gradient model

Reference

1. N. Lourenço, N. Horta, in *GENOM-POF: Multi-Objective Evolutionary Synthesis of Analog ICs with Corners Validation*. GECCO' 12: Proceedings of the 14th International Conference on Genetic and Evolutionary Computation Conference, July 2012, pp. 1119–1126